Springer Theses

Recognizing Outstanding Ph.D. Research

Aims and Scope

The series "Springer Theses" brings together a selection of the very best Ph.D. theses from around the world and across the physical sciences. Nominated and endorsed by two recognized specialists, each published volume has been selected for its scientific excellence and the high impact of its contents for the pertinent field of research. For greater accessibility to non-specialists, the published versions include an extended introduction, as well as a foreword by the student's supervisor explaining the special relevance of the work for the field. As a whole, the series will provide a valuable resource both for newcomers to the research fields described, and for other scientists seeking detailed background information on special questions. Finally, it provides an accredited documentation of the valuable contributions made by today's younger generation of scientists.

Theses are accepted into the series by invited nomination only and must fulfill all of the following criteria

- They must be written in good English.
- The topic should fall within the confines of Chemistry, Physics, Earth Sciences, Engineering and related interdisciplinary fields such as Materials, Nanoscience, Chemical Engineering, Complex Systems and Biophysics.
- The work reported in the thesis must represent a significant scientific advance.
- If the thesis includes previously published material, permission to reproduce this must be gained from the respective copyright holder.
- They must have been examined and passed during the 12 months prior to nomination.
- Each thesis should include a foreword by the supervisor outlining the significance of its content.
- The theses should have a clearly defined structure including an introduction accessible to scientists not expert in that particular field.

More information about this series at http://www.springer.com/series/8790

Thomas Klein Kvorning

Topological Quantum Matter

A Field Theoretical Perspective

Doctoral Thesis accepted by
Stockholm University, Stockholm, Sweden

Springer

Author
Dr. Thomas Klein Kvorning
Department of theoretical physics
Royal Institute of Technology
Stockholm, Sweden

Supervisor
Prof. Thors Hans Hansson
Stockholm University
Stockholm, Sweden

ISSN 2190-5053 ISSN 2190-5061 (electronic)
Springer Theses
ISBN 978-3-030-07246-9 ISBN 978-3-319-96764-6 (eBook)
https://doi.org/10.1007/978-3-319-96764-6

This Springer imprint is published by the registered company Springer Nature Switzerland AG
The registered company address is: Gewerbestrasse 11, 6330 Cham, Switzerland

Supervisor's Foreword

The notion of topological states of matter dates back to the discovery of the quantum Hall liquids in the early 1980s. The term topological order was introduced by X.-G. Wen in the early 1990s to characterize those topological phases that support quasiparticles with fractional quantum numbers, the archetypical example being the anyons in the fractional quantum Hall liquids, that carry a fraction of the unit electric charge. The existence of these exotic particles was predicted by R. Laughlin, and the fractional charge was later observed in several experiments. Another important property of topologically ordered states is the gapless modes that are present at the boundaries of a sample. If the bulk is two dimensional, these states are found at the edges, and in the quantum Hall effect, it is these edge modes that are responsible for the quantized current response.

In 2007, the first topological insulators were discovered. These states are quite similar to the integer quantum Hall liquids in that they do not support fractional quasiparticles, but just as the topologically ordered states they support gapless modes at the boundaries between states with different topological properties. These topological, but not topologically ordered, phases are now referred to as symmetry protected, phases, since they are distinct only as long as certain symmetries are present. As opposed to the topologically ordered states, which are always interacting, the essence of a symmetry protected topological state can be captured by a model of noninteracting fermions.

The symmetry protected topological phases do not only include topological insulators but also certain superconductors, as long as one neglects the fluctuations in the electromagnetic field. A particularly interesting class of topological superconductors are those referred to as chiral. In these super conductors the Cooper pairs, which are loosely bound states of two electrons, carry an orbital angular momentum, which in a two-dimensional system to good approximation can be considered as perpendicular to the surface where the electrons move. If this surface is curved, one realizes the archetypical situation for a quantum mechanical Berry phase, which turns out to have interesting consequences for the electromagnetic response. The situation becomes even more interesting if one specializes in topological chiral superconductors where the electron pairing takes place in an odd angular momentum channel such as p or f. In this case, the edge states are very peculiar—they are so-called Majorana fermions which in a sense can be thought of as half fermions. These Majorana modes also occur at vortices in the bulk of the superconductor, and are of great current interest since they can in principle be used as robust carriers of quantum information.

It was later realized that topological states of matter were already known prior to the proper recognition of their special properties. It was shown that ordinary superconductors are topologically ordered when one includes the effects of electromagnetic fluctuations, and also that certain spin-chains, introduced by F.D.M. Haldane in 1983, are in fact examples of symmetry protected topological states.

All of these very topical subjects—topologically ordered, and symmetry protected topological phases, superconductors and Majorana modes—are covered in the Ph.D. thesis, Topological quantum matter—a field theoretical perspective, written by Dr. Thomas Klein Kvorning. I believe that the way this thesis is written makes it suitable as an introduction to field for those who strive for a deeper understanding of the theoretical description of topological phases. Many elementary treatments focus on specific lattice Hamiltonians representative of some specific symmetry protected topological states. Given such a Hamiltonian, one can do computer simulations and calculate, for instance, various properties of the edge states. This is often a very fruitful endeavor since, by choosing realistic lattices and geometries, there is often a direct connection to experiments. However, to get a deeper perspective on topological matter, computations are not enough. One must understand what characteristics are "topological" and which are not. To do so, the notion of topological quantum numbers and topological field theory is essential. The latter are special quantum field theories that are insensitive to details of the interaction between the electrons, and which encode all the topological information. Thomas' thesis provides an excellent introduction to the field theoretic description of various topological phases, and in appendices, he summarizes the required mathematics to make the text self-contained. It should be accessible to students with basic knowledge of quantum field theory and condensed matter physics, and could also be used by more senior condensed matter physicists who want a concise introduction to field theory methods for topological matter.

I write this as Dr. Klein Kvorning's supervisor, and I take the opportunity to say that it has been a pleasure to work closely with Thomas, and I have also very much enjoyed the collaboration we both had with Prof. Cristiane Morias Smith and Dr. Anton Quelle at the University of Utrecht.

Stockholm, Sweden
June 2018

Prof. Thors Hans Hansson

Acknowledgements

First I would like to thank my supervisor. Hans, you have taught me very much about physics, helped me when I needed, and given me much encouragement. But above that, you have been a good friend. I have much enjoyed talking to you and getting advice, also on other topics than physics. The many times you invited me to your home are all very fond memories. I would also like to mention that I am glad that you taught me, by explicit example, that to avoid giving your coauthors headache, you should not insist on disregarding Latex standards.

I would also like to thank my coauthors, Parameswaran, Sreejith, Shinsei, AtMa, Eduardo, Anton, and Cristiane. It has been an absolute pleasure to work with all of you!

I am a student at Stockholm University, but I have also been part of the condensed matter group at UIUC. I had a great experience in Illinois, thanks to all of you that I met to work and socialize with—not least you, AtMa, Xueda, and Apoorv. Thank you!

The atmosphere in the research group and the theoretical physics community here in Stockholm has been fantastic. This is thanks to many, definitely including you, Fawad, Eddy, Supriya, and Maria. Jonas, you have also contributed: you bring the best gossip in the corridor, and I have learnt much from and enjoyed all discussions we have had (excluding discussions concerning software installation). Fernanda, Micke, Christian, and Flore, with you I have discussed physics, traveled to conferences and studied quantum field theory till early hours in the morning—I could not have wished for better fellow Ph.D. students!

Samuel, we have been close friends since the beginning of our physics studies. Our discussions have contributed much to my knowledge, both in physics and elsewhere.

Emma J., you and I have been friends long before we shared office, but during our studies you have become family. You are the sunshine of my life, sometimes even outside the office.

Emma W., during my Ph.D. studies, you have become one of my closest friends. You have also helped me with many things related to my studies. One of those things is that you have corrected my grammar and spelling throughout this thesis, and done so on short notice whenever I needed! You have also supported me and encouraged me—always! I remember when I started as a Ph.D. student: then you invited me to your home, we baked bread, and you gave a pep talk.

Sören, I do not know where to begin. I could devote an entire chapter to you. You are truly a great friend. You have helped me so much and made me overcome so many frustrating moments. I have learnt extremely much from the discussions we have had, about physics and philosophy but also the importance of having a fruit break everyday at three o'clock. Also, not to forget, you have drawn more than half of all the figures in this thesis!

There are also many people who have not been directly connected to my doctoral studies, but who made the thesis possible. Mom and dad, thank you for giving me so much support even when you have many troubles of your own. No matter how hard I try, I cannot imagine any better parents! Thank you Mikael for your support and all the discussions we had growing up—you developed my arguing skills. My wider family; cousins, uncles, aunts, and grandparents: not all of you are here, but all of you were there when I started my Ph.D. studies and you have all, always, encouraged and motivated me, which has been a great help.

Bisse, you are also family! You have been with me my entire adult life, and have always been an amazing friend. I can with certainty say that I would not have finished this thesis without you.

I also want to thank all my other friends—so many of you have been a wonderful support! Lately, I have not given many of you the attention you deserve—something that I hope will change in the near future.

Finally, I would like to say that I dreaded writing the main text of this thesis. I usually get in a very bad mood by deadlines and stress. At the same time, it has been a very turbulent period in my life, with many opportunities to have my spirit destroyed. That has not happened. This year has been one of the absolutely happiest in my entire life, and that is because of you, Josefin. You have given me so much support and helped me overcome the problems I encountered.

Contents

Chapter 1
Introduction

As a child I was fascinated with the fact that you could not leave a bucket filled with water outside on a cold winter evening and expect to find it the same in the morning. In the evening the bucket was full of water—a liquid, which can be stirred and poured—but in the morning the bucket is filled with a material with almost opposite properties: hard, rigid and stiff. Growing up I learnt what kind of phenomena I had witnessed, *a phase transition*: a drastic change of a state of matter.

That properties of matter sometimes have to undergo a dramatic change, and cannot change continuously, not only fascinated me as a child—it still does today. It is also the main topic of this thesis. But the phases I will discuss here are *topological quantum phases*, and they are even more fascinating than the phases of water. I will begin by giving some flavor and an intuition for the subject with as little use of technical mathematical language as possible. To best appreciate the topological quantum matter in full I will of course make use of a considerable amount of mathematical language, but that will be saved for the coming chapters.

Properties of matter will vary when parameters, such as pressure or temperature, is varied. Usually these changes happen gradually—a small perturbation in the parameters leads to a small change in the properties of matter. However, certain properties cannot change gradually. When varying the parameters these properties remain constant, until a critical point is reached, when even a tiny perturbation of the parameters lead to a dramatic change. Water and ice is distinguished by such a property that cannot change gradually. In that case it is the periodicity of ice that is not present in water.

The configuration of water molecules is determined by laws of nature that possess translation symmetry. This means that the laws do not distinguish different points in space; they are all equivalent. The same is true for water (i.e., the phase of matter): water is homogeneous, and different points or directions in space cannot be distinguished. This is not true for ice. If you study ice with a sufficiently good microscope you would discover that the molecules are ordered in a lattice, which look very different *at* the lattice points as opposed to *in between* the lattice points. This periodic

T. Klein Kvorning, *Topological Quantum Matter*, Springer Theses,
https://doi.org/10.1007/978-3-319-96764-6_1

ordering is an example of *spontaneous symmetry breaking*. Ice is a collection of water molecules and is thus described by laws of nature with translation symmetry, a symmetry which the ice does not posses. Ice has thus spontaneously (i.e., not by an external force) broken a symmetry. This is the most common way phases are distinguished, and that understanding dates back to the work of Lev Landau in the 1930s, Ref. [1]. However, the phases I will discuss in this thesis, *topological matter*, are distinguished in a drastically different manner.

1.1 Entanglement

Different topological quantum phases do not differ by anything that directly could be observed in a microscope, whatever its resoluting power. In terms of the Landau-characterisation of phases they are all liquids, in the sense that no spontaneous symmetry breaking have occurred—the different phases do not have different symmetries. What can then be the difference? The answer lies in structures of the quantum entanglement of the states. Or in other words, non-separable information in the states. To understand what this means we must go through some fundamental facts concerning quantum mechanics.

Everything around us is described by quantum mechanics, and this has some remarkable consequences. For almost all practical purposes one can divide any physical object into parts. Consider a brick wall, and let's say that you know everything there is to know about every brick—every possible property, down to the last atom. Then one would think that you also would know everything about the entire wall. After all, it is nothing but its parts? Yes it is, actually! There is also non-separable information: information that does not exist in the sum of the parts, but which can only be attributed to the brick wall as a whole. It is such non-local properties that distinguish different topological phases of matter.

In general, there is much more spread out non-separable information than there is separable, i.e., the sum of all local information. Even so, everyone knows from experience that most things around us can be treated by dividing them into parts. The reason for this is another fundamental property of our world: it is not possible to directly observe non-separable information. Everything you can see in a microscope, with the naked eye, or by any other observation means (no matter how advanced) is always separable information. The non-separable information only manifests itself indirectly through interaction with separable degrees of freedom.

In a typical system, the effect on the local degrees of freedom, from the interaction with non-sepparable degrees of freedom, will be random and will just be a part of the thermal fluctuations. We would like to get away from this quite generic situation where the non-sepperable information is uninteresting. One way to accomplish this is to consider systems where there is minimal energy, Δ, required to affect certain properties, e.g., that no charge will flow unless enough energy is provided, i.e., the system is an insulator. (To avoid complicating things and keep the discussion focused on the topics characteristic to topological quantum matter, I will simply assume that when the system is at zero temperature i.e., when it is in the ground state then there is an energy gap, Δ, common to *all* excitations.) If the Boltzmann constant, k_B, times the temperature, T, is much lower than this energy gap, $\Delta \gg k_B T$, then the thermal fluctuations effectively disappear and the possibility for topological matter emerges.

For most topological matter the mentioned energy gap is very small, which means that they are only realized at very low temperatures. But there are also situations where the gap is large. The most extreme example is graphene in a very strong magnetic field. Then the integer quantum Hall effect (a topological quantum matter phase I will discuss later) can be observed at room temperature.

We have now come to the question: How does the non-separable information manifest itself in topological quantum matter? This question has rather different answers depending on which type of topological matter you are considering. The topological matter of interest in this thesis can be divided into two categories: matter with *topological order* (TO) and *symmetry protected topological* (SPT) matter. Let us begin by considering topologically ordered matter, and specifically topologically ordered matter in two dimensions. That is, I assume that the matter is made up of building blocks confined to sheets so thin that the building blocks only can move in two perpendicular directions.

1.2 Topological Interactions

The key aspect of topologically ordered matter is the particles that the matter allow. When I say *particle* I do not just mean elementary particles, but also particles that only can occur in certain phases of matter (often referred to as quasi-particles). Vacuum is for condensed matter physicists a phase of matter, as any else. It is totally homogeneous everywhere, but in this homogeneous background you can find small regions containing something else: it could for instance be photons or electrons. It is these types of possible local alterations to pure vacuum that we call elementary particles. In the same way, other phases also allow for particles, and this can be defined even if they are not homogeneous, as long as there is a well defined background.

Consider ice for example. It has a lattice structure which makes a well defined background. One can therefor, in the same way as with the vacuum, tell if there is a particle, that is a local alteration of the periodic lattice. The particles in ice are, among others, phonons (sound quanta) and photons.

In topologically ordered matter there are particles called *anyons* with a very special property—their interaction. At first sight these particles do not seem to interact at all, at least not on long distances. But with a closer look, one will realize that there is a subtle form of interaction: *topological interaction*. The state of the system will depend not only on how the individual particles move, but also on how the particles have *braided* (i.e., encircled each other), see Fig. 1.1. This is something that cannot be associated with any particular particle; it can only be associated with the particles as a collective, which means that it is some kind of interaction.

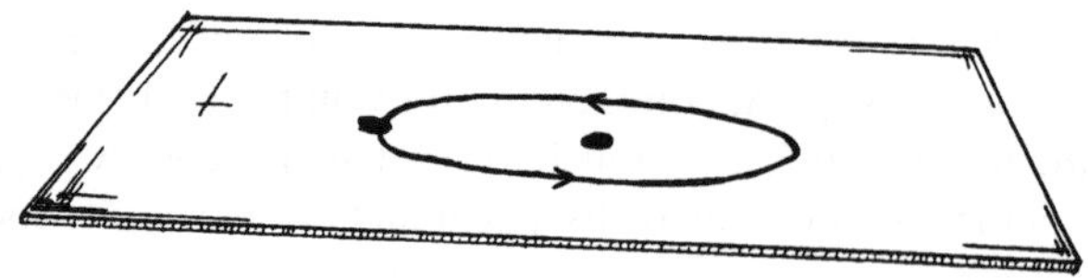

Fig. 1.1 A particle encircling another

There are two different classes of topological ordered matter: one which supports only *Abelian* anyons and one which supports *non-Abelian* anyons. The Abelian anyons have the property that if you have some complicated process where several different particles have braided, the order in which the encirclings have occurred does not matter. For the Abelian anyons the effects of the interaction is very subtle—it is responsible for something called a Berry phases of the state. Later I will discuss this concept and you will see that the Berry phases connected with the topological interactions can have far reaching consequences. But here I will instead focus on the non-Abelian anyons where the effects of the interaction is more direct.

Assume that you locally in a system create a particle and its anti-particle, close together. The system could be vacuum and the created excitation an electron-positron pair. You could then take the electron and positron far away from each other and the state could then not return to the pure vacuum state, without moving them back together. However, if you where to move them back together you could always make them annihilate, no matter what happened to them when they where separated. For non-Abelian anyons the situation is quite different. If you create one anyon together with its anti-anyon and bring the anyon and anti-anyon far away from each other, you can make them annihilate when you bring them back together. But if you first let the anyon braid with some other anyon in the system, it is not sure that the anyon can annihilate against the anti-anyon any more.

As long as the anyons are far away from each other, braiding will not lead to any measurable difference of the system, even if the property of them being able to annihilate each other is altered. However, when an anyon and an anti-anyon get close to each other the difference can of course be measured: in one case they will annihilate and in the other they will not. How can this be, first there seemed to be no information of how the particles braided, and then suddenly there was? The reason is that, when the anyons are far way from each other the information of braiding is stored in non-separable degrees of freedom which become local when the anyons are brought close together. This controlled interplay between local and non-separable degrees of freedom is the hallmark of topological order.

The result of what happens when to particles fuse together cannot vary continuously (consider e.g., the difference between if they can annihilate and if they cannot) and it depends on the topological interaction in the system. There thus exist distinct phases of matter characterized by their topological interactions.

At this point it might have occurred to you to ask why the interactions I have discussed are called topological. Topological properties of a system are generally defined as properties that cannot change gradually, i.e., they do not depend on any change no matter how big, as long as it can be decomposed into several smaller

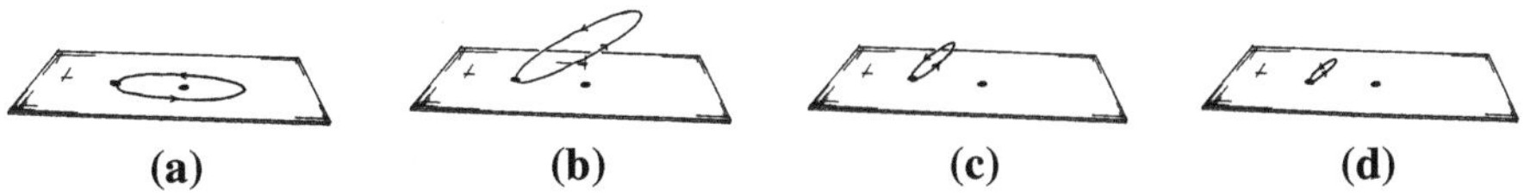

Fig. 1.2 In three dimension you can deform a process where one particle encircle another to a process where they do not

changes. The outcome when the particles are fused (i.e., the braiding of the particles that have occurred) does not depend on the distance between any pair of particles, which usually is the case for interaction. It actually does not depend on much at all. As long as you only do changes gradually, and avoid that particles end up at the same point, the outcome does not depend on the path of the particles at all. The interaction thus only depends on properties of the paths that cannot vary continuously: in which order and in which direction, the particles have encircled each other.

You might now have figured out why I have limited my self to states in two dimension. In two dimensions it can be distinguished whether particles have encircled each other or not, and topological interaction of them is possible. In Fig. 1.2 you can see that in three dimensions one can always continuously deform a process where one particle encircle another, to a process where they do not. There *is*, however, also topologically ordered matter in three dimensions, but then the different phases distinguish them selves with different topological interaction involving strings instead of just point particles.

1.3 Topological Response

As you now have seen the topologically ordered phases have the astonishing property that they allow for some of the non-separable information to be accessible by interacting with local degrees of matter through braiding of particles.

The other topological phases to be considered, in this thesis, goes under the name symmetry protected topological (SPT) phases. As the name suggests, these phases are protected by a symmetry in the sense that two phases are distinct only if a certain symmetry is preserved. By breaking that symmetry one can continuously interpolate between two different symmetry protected phases, and they should thus only be considered different phases in presence of the symmetry.

In these phases one cannot access non-separable degrees of freedom as in topologically ordered phases, but non-separable information still plays an important role. Here these degrees of freedom act in a way which makes certain macroscopic measurable quantities exactly quantized. Let us consider the most famous example—that the Hall-resistance in certain circumstances is quantized in two dimensions. In that case the symmetry which protects the phase is the charge conservation symmetry.

Normally the resistance R of a system, that is the proportionality constant between a current I through the system and the voltage U across the system,

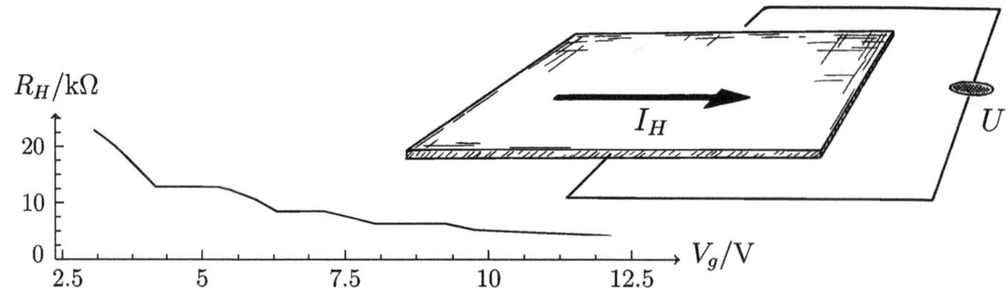

Fig. 1.3 On the right there is a schematic illustration of the quantum Hall experiment. The graph shows measurement data from the article where the quantum Hall effect was first reported, by von Klitzing et al. The electron density scales with V_g, on the horizontal axis. When it is changed the system undergoes several phase transitions. The horizontal plateaus correspond to situations where the $k_B T$ is much smaller than the energy gap of the system. By a closer inspection it can be seen that the plateaus occur at a Hall resistance, R_H, that equals $\frac{h}{e^2}$ divided by an integer

$$U = RI \ , \tag{1.1}$$

depends on practically an infinite number of details of the system. The Hall-resistance in two dimensions is different. Let us first recall what it is. Say that you have a two dimensional, rectangular system and that there is a voltage U, between two of the edges of the rectangle (see the righthand side of Fig. 1.3). In most materials this would imply that there is a current between the edges with a voltage difference, but for certain matter where the Hall-resistance is non-zero there is also a current I_H between the other edges. The proportionality constant between the current I_H and the voltage U is precisely the Hall-resistance R_H,

$$U = R_H I_H \ . \tag{1.2}$$

If the condition $\Delta \gg k_B T$, mentioned above, is full filled, then the Hall-resistance is quantized. It is to an immense precision the reciprocal of an integer times Plank's constant divided by the elementary charge squared, h/e^2. Since an integer cannot change continuously a phase transition is required between every state with different Hall resistance, see Fig. 1.3. This is remarkable! Remember that I am discussing macroscopic systems which depend on a practically infinite number of parameters. Even so, if you keep a constant voltage the current will be *exactly* the same between samples that can vary extensively. If the current for some reason is blocked in one part of the rectangle the current would increase exactly in the right amount in another part of the rectangle, to compensate. How can this be?

This is one of few precise experimental facts that have a simple analytically derivation, directly from the microscopic laws of nature.

I will go through this proof in Chap. 3 and I hope you bare with me and read until then since it is very beautiful, indeed. In this introductory text I cannot say much without the discussion becoming too technical. What I can reveal is that when the temperature is small enough, $\Delta \gg k_B T$, then the state of the system defines a geometry. Not a surface in physical space, but a higher dimensional geometry on

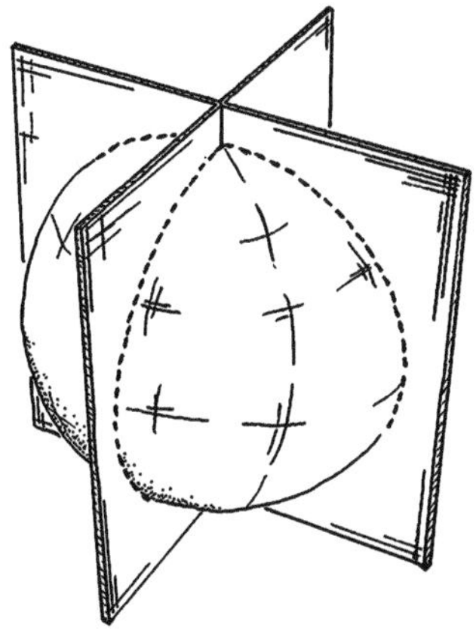

Fig. 1.4 The surface mentioned in the text is here the sphere and the point is where the dashed lines meet. The two planes are two examples of planes that contain the normal of the surface at the point

an abstract space related to the quantum mechanical formalism. It turns out that the quantum Hall resistance is a property that seemingly depends on a lot of details about this geometry, but that actually only depends on its overall shape—its topology.

How this happens is analogous to a problem that is much easier to visualize and that provides the right intuition. Namely, how the total Gauss curvature of a surface depends on its shape.

So, what is Gauss curvature? Take a surface, and draw a plane containing the normal to the surface at some point p, see Fig. 1.4. The intersection between the surface and the plane is a curve, and close to the point p you can approximate this curve by a circle (in Fig. 1.4 the curve is a circle so the approximation is, in fact, exact). For a general surface the size of the circle will vary as you rotate the plane, but there is a certain position of the plane which gives a circle with maximal radius—I call it R_1. There is also another plane, containing the same normal, which instead gives the smallest circle, and that I call R_2. If the smallest and the largest circle are such that if you try to shape your hands like the circles and the hands lie with one palm on top of the other hands back (like the circles in Fig. 1.4) then the curvature K at the point p is given by

$$K = \frac{1}{R_1 R_2} . \tag{1.3}$$

If the hands instead would lie back against back, or palm against palm, then the curvature would be given by

$$K = -\frac{1}{R_1 R_2} . \tag{1.4}$$

If you integrate the curvature over all points on our surface you get the total curvature of the surface. If you do this for several different surfaces you will discover something remarkable. The total curvature is *always*

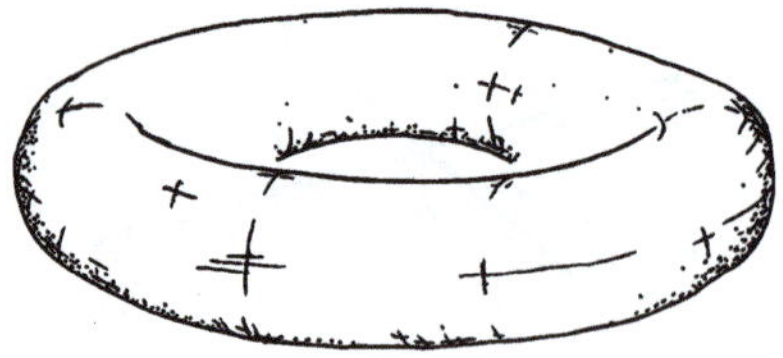

Fig. 1.5 A torus

$$4\pi(1-h)\ , \tag{1.5}$$

where h is the number of holes through the surface (a sphere has zero holes, a torus, see Fig. 1.5, has one hole, etc.).

This is quite remarkable. It means that if you try to flatten out the sphere in some region it will exactly compensate by curving a bit more somewhere else. If you form a dent in the shape of a small half sphere you necessarily have to compensate by having some negative curvature somewhere else. It turns out that the quantum Hall conductance works exactly in the same way. Instead of being a total curvature it is a more abstract concept called total Berry curvature, and instead of the geometry being a physical surface it is an abstract geometry related to the quantum mechanical formalism, but otherwise the situations are very similar.

For the quantum Hall effect I discussed how electric charge responded to a voltage. But this is just a particular example of a much more general phenomenon, of how some conserved quantities topologically respond to external perturbations. Another example is how a magnetic field responds to bending or straining a system. Magnetic flux is always conserved, but typically there is no energy gap to moving flux, except in superconductors. Considering how flux responds to bending a material and creating curvature one will see that there also is a quantized response. This is a very new discovery called the *geometric Meissner effect*, see Ref. [2], which I will discuss in the last chapter. Among other things, it means that certain superconductors will spontaneously produce magnetic fields when curved, see Fig. 1.6.

Fig. 1.6 Magnetic field lines for a superconductor with a geometric Meissner effect. *Image by S. Holst*

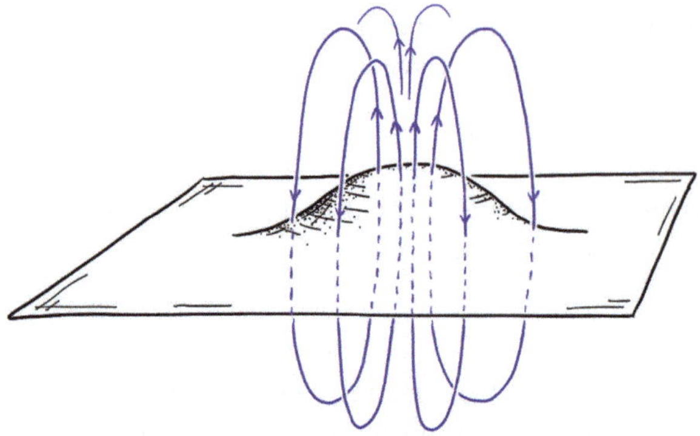

1.4 Outline

I begin, in the next chapter by studying topological interactions in general and introduce topological field theory and see how it relates to topological interactions.

Then, in Chap. 3, I will move on to study SPT phases and introduce their hallmark namely non-trivial topological response. I will mainly consider systems with $U(1)$ symmetry (i.e., ordinary charge conservation) and investigate the topological response to probing it. I will also discuss the connection between topological response and topological order.

This naturally brings us to the topic of geometric response and more particularly geometric response of chiral superconductors which is the subject of the last part of Chap. 3.

The final Chap. 4 is devoted to superconductors, and primarily chiral superconductors. These will be a concrete example of both the topological order discussed in Chap. 2 and topological response discussed in Chap. 4. I will discuss the non-Abelian topological order in chiral p-wave superconductors as well as the recently proposed geometric Meissner effect—how curvature can produce spontaneous magnetic fields in chiral superconductors.

As a final note. The subject of topological quantum matter is discussed in different mathematical language. When choosing which to use there is a trade off between the language that in a least involved and in a most direct way reveal the physics unique to topological quantum matter and a language that is familiar to more physicists but for this purpose will be more involved. I choose to use the language I find most direct and what I a few years ago would have appreciated the most. I will below give a short motivation for this and in the appendices you can find definitions and short explanations for the mathematics used.

1.5 Notations and Conventions for Topological Matter

In Appendix A and B I introduce much of the mathematical language I will use in this thesis. But before you go there (or to the next chapter for that matter), I would like to motivate the choice of mathematical language used in this thesis.

In most areas of physics one deals with systems where the size of objects are of great importance—i.e., length, area, volume and so forth. This information is captured in a metric which often is seen as so natural that it is only implicitly present in the notation. This thesis does concern properties of matter that are connected to geometry. However, those properties do not depend on a background metric (or other geometric structures). Rather, what I am after are *topological* properties, meaning that the most familiar geometrical language is not necessarily the most transparent. I will make use of the language of exterior calculus and differential forms, and a non-standard normalization of the electromagnetic fields. Let us start by an example that hopefully will make its usefulness clear.

Usually one encodes the information of where charge is located and where current flows by a current space-time density vector j^μ. But on its own you cannot use this information to know how much charge there is in a spatial region R, or how much charge that, has flown through a surface Γ during some time-interval. You need more information, namely how big the regions are. The vector j^μ is useless without knowing the (sometimes implicit) space-time metric which is used to define it. For example, if you have a parametrization $(s,t) \rightarrow X(s,t) \doteq (x^1(s,t), x^2(s,t), \ldots)$ of the surface R, the charge is given by

$$Q = \iint dsdt\, \sqrt{|g|}\varepsilon_{\mu\nu\sigma} j^\sigma \frac{dx^\mu}{ds}\frac{dx^\nu}{dt}\,, \tag{1.6}$$

where $\doteq$ denote "represented by", ε is the totally antisymmetric tensor, $|g|$ is the absolute value of the determinant of the metric and Einstein summation convention is assumed (I now assume $2+1\,\mathcal{D}$).[1] The charge seems to depend explicitly on the metric, but the information of how much charge there is in a volume is independent of its size, or any other metric property. This information can be encoded without reference to a metric if one uses an object called the current form $\mathcal{J}$. In $2+1\mathcal{D}$ it is a two-form (see Sect. A.1 of the appendix), i.e., an object which assigns a value to each surface. This value is the charge that has passed through some line in the case of a non-spatial—and the charge on some surface in the case of a spatial—surface.

Written in this way, you can also see why, e.g., the Hall-conductance in $2+1\mathcal{D}$ is a topological response. It is nothing but a proportionality between the current two-form and the electromagnetic field-strength tensor,

$$\mathcal{J}_q = \frac{\sigma_H}{2\pi} F\,. \tag{1.7}$$

Without reference to any size it thus encode a proportionality between the total charge on—and total flux through—each space-time surface.

In Appendix A.1 I will introduce the notion of differential forms and exterior calculus that I now touched upon. There you will also find conventions for normalization. I should mention that the electromagnetic field strength F and the electromagnetic vector potential A is measured in units of $\phi_0/2\pi$, where ϕ_0 is the flux quanta, $\phi_0 = h/e$.

In Appendix B you can find a collection of technical definitions of some the mathematical terms used in this thesis.

[1] I use lower case calligraphic $\mathscr{d}$ to denote the spatial dimensions of a system and upper case $\mathcal{D}$ to denote the space-time dimensions of the same. Furthermore, e.g., $2\mathscr{d}$ will denote a system with two spatial dimensions while a e.g., $3\mathcal{D}$ denotes a system with $2+1$ dimensions.

References

1. L.D. Landau, Phys. Z. Sowjet. **11**, 26 (1937), https://www.ujp.bitp.kiev.ua/files/journals/53/si/53SI08p.pdf
2. T. Kvorning, T.H. Hansson, A. Quelle, C.M. Smith, Phys. Rev. Lett. **120**, 217002 (2018). https://doi.org/10.1103/PhysRevLett.120.217002

Chapter 2
Anyons and Topological Order

Topologically ordered phases differ by the topological interaction between the particles they support and you will understand why when you get to Sect. 2.2 of this chapter. However, I will begin this chapter with a more familiar topological interaction, the distinction between bosons and fermions. Attacking this problem from the right angle leads to the notion of particles with other topological interactions—the Abelian anyons. Here you will also see how topological interactions can be understood as an exchange of a Chern-Simons (CS) gauge boson.

Finally I will discuss topological interactions, and thus also topological order in general, and introduce the notion of non-Abelian anyons. The latter will reappear in Chap. 4, where chiral p-wave SC's are studied.

2.1 Bosons, Fermions and Abelian Anyons

One of the first things one learns when dealing with many-body quantum mechanics is that there are two distinct classes of identical particles: bosons and fermions. You can distinguish them, at least in a thought experiment, where two particles are moved around each other and end up in an configuration where they have changed place. Since they are indistinguishable this does not change the state, all you get is a phase shift of the wave function. If the particles are kept far enough apart there will be no interaction between them (if they are bosons) and the phase shift is a sum of contributions from each of the particles involved in the exchange.

On the other hand, if they are fermions there will be an extra minus sign that cannot be contributed to either one of the two particles, i.e., it is an interaction effect. Since it is a very special type of interaction, it does not depend on the distance between the particles, only on the topology of the paths they take. It is however usually not called an interaction; instead one usually talks about the *statistics* of the particles.

T. Klein Kvorning, *Topological Quantum Matter*, Springer Theses,
https://doi.org/10.1007/978-3-319-96764-6_2

The minus sign obtained because of an exchange of fermions is the simplest type of topological interaction and in three dimensions and higher it is the only type of topological interaction possible between point particles. But in 1977, Jan Myrheim and Jon-Magne Leinaas, Ref. [1], realized that for two dimensional systems the situation do not need to be quite that simple. In two dimensions anyons—particles with other topological interactions—are possible.

At their discovery anyons where a purely theoretical construct, but things changed with the discovery of the fractional quantum Hall effect, Ref. [2]. In 1984 Bert Halperin, Ref. [3], made an argument for why the then recently discovered FQH state support anyons, something that has been put on a very strong theoretical footing.[1] In 1991, see Refs. [6, 7], it was suggested that also non-Abelian statistics could be realized in some FQH states. And, more interestingly for this thesis, also in chiral p-wave superconductors, see Refs. [8–11].

The presence of anyons is the hallmark of topologically ordered states. So, in this chapter I begin by a carefully exposé of the topological interaction and see why two spatial dimensions is special, i.e., why fermions are the only non-trivial case in dimensions other than two. But first we warm up and start with a simpler setting, namely, single particle physics.

2.1.1 A Single Particle on An Annulus

The kinetic energy of a state is related to how much the wave functions varies. The modulus squared of the wave function is the probability density, so for that there is a well defined notion of how much it varies. For the phase, on the other hand, there is not. There is a priori no notion of how to compare the phase of a wave function at two different points. To define the change $\delta\psi$ of a wave function ψ in some direction, i.e., along short path Γ, you need a $U(1)$ connection A,

$$\delta\psi = \int_\Gamma (d + iA)\,\psi\,, \tag{2.1}$$

and then you can define a kinetic term. If ψ describes a charged particle, A is just the potential for the magnetic field,

$$\mathcal{B} = dA\,, \tag{2.2}$$

[1] A very important part of this is when, in Ref. [4], it was shown that the model state proposed by Robert Laughlin, see Ref. [5], support Abelian anyons. Maybe even more important is, as you will see later, that a fractional Hall conductance can only occur if there is a topological torus degeneracy, which is directly tied to the presence of anyons.

here $\mathcal{B}$ denotes the magnetic field two-form and B denotes the magnetic field scalar (see Example 16 in Appendix B). In quantum mechanics, as opposed to the classical case, it is not just the field strength $\mathcal{B}$ that is important: in a classical theory, A is only defined up to a closed form that is a form a with $da = 0$. However, in quantum mechanics closed but non-exact $U(1)$ connections make a physical difference. An exact form is a form that can be written as $a = d\lambda$, and the notion *exact* $U(1)$ *form*, which I also will refer to, is a form that can be written as $a = i\xi^* d\xi$, where ξ is a complex function with unit modulus.

A requirement for there to be any non-trivial, closed non-exact forms is that there is at least one closed curve in the configuration space, $\mathcal{M}$, that cannot be contracted to a point. (In this thesis I always implicitly assume and orientation for all manifold and sub manifolds, i.e., curves, surfaces, etc.) So, I now take $\mathcal{M}$ to be a simple example of such a manifold–a plane with a hole cut out. In that case one knows, physically, what the closed non-exact $U(1)$ form corresponds to, namely magnetic flux passing through the hole.

Let us consider the simplest non-trivial case, that is, to consider A to be closed and assume that $e^{i\int_\Gamma A} = -1$, where Γ is any curve encircling the hole. I will employ a mathematical trick: I do not consider the wave functions to be functions on $\mathcal{M}$, but to be part of a subclass of functions on the double cover ($\tilde{\mathcal{M}}$) of $\mathcal{M}$, described in Fig. 2.1. When viewing functions on $\mathcal{M}$ as functions on $\tilde{\mathcal{M}}$, the covariant derivative is

$$d + i\tilde{A} \; , \tag{2.3}$$

where $\tilde{A}$ is a one-form on $\tilde{\mathcal{M}}$ that defines A in the following sense: Take any curve Γ and pair it up with a curve $\tilde{\Gamma}$ which have Γ as image under the mapping χ, described in Fig. 2.1. Then you get $e^{i\int_\Gamma A}$, from $\tilde{A}$, as

$$e^{i\int_\Gamma A} \equiv e^{i\int_{\tilde{\Gamma}} \tilde{A}} \; . \tag{2.4}$$

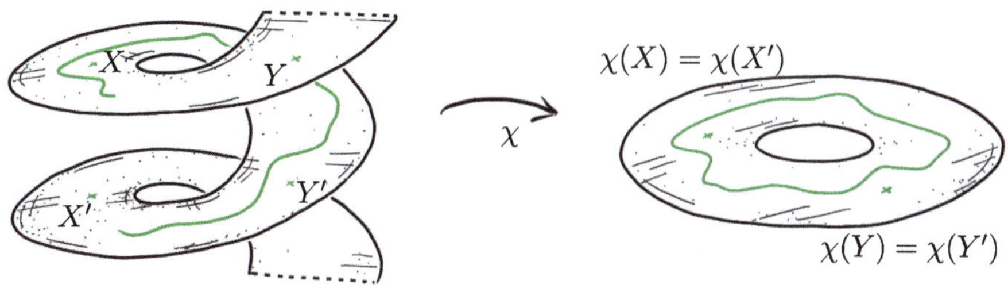

Fig. 2.1 $\tilde{\mathcal{M}}$ (left) is up to a line (the points on the dashed line should be pairwise identified) two copies of $\mathcal{M}$ (right). By the mapping χ, depicted in the figure, closed curves in $\mathcal{M}$, winding once around the hole, are images of open curves in $\tilde{\mathcal{M}}$. In the figure you can see pair of points (X, X') which are points that get mapped to the same point under χ. *Figure by S. Holst*

For all curves in $\tilde{\mathcal{M}}$, the integral $e^{i\int_{\tilde{\Gamma}}\tilde{A}}$ only depends on the end points of $\tilde{\Gamma}$, which means that $\tilde{A}$ is an exact $U(1)$ form. Hence, $e^{i\int_{X_0}^{X}\tilde{A}}$ is a well defined function, where $\int_{X_0}^{X}\tilde{A}$ refers to the line integral over any curve starting at X_0 and ending at X. You can thus do the gauge transformation

$$\psi(X) \rightarrow \psi(X)e^{i\int_{X_0}^{X}\tilde{A}}, \tag{2.5}$$

which results in $\tilde{A} = 0$. However, now the functions cannot be interpreted as functions on $\mathcal{M}$. For that to be the case they need to be in the even subclass of functions on $\tilde{\mathcal{M}}$, but ψ is now in the *odd subclass*. The even subclass of functions on $\tilde{\mathcal{M}}$ consists of functions that take the same value on both points in each pair $(X, X' \in \tilde{\mathcal{M}})$ that get mapped to the same point in $\mathcal{M}$ under χ, i.e., $\psi(X) = \psi(X')$. Analogously, the odd subclass of functions are the functions with the property $\psi(X) = -\psi(X')$.

One can thus conclude that the closed but non-exact form A could be removed by a *singular gauge transformation*. That is, one can consider the wave functions to be functions on a multi-cover of our surface, and consider a certain subclass of functions there.

In the same way, any closed form can be gauged away. However, one can then no longer consider the functions to be functions on our original surface $\mathcal{M}$, since the function is not well defined. In the case discussed here there are two different function-values for each point in $\mathcal{M}$, which are related by a minus sign, and in general by $e^{i\int A}$ over a closed non-trivial curve. I will refer to this factor, between different function values, as the *monodromy* of the wave function.

2.1.2 Indistinguishable Particles and the Braid Group

Now let us consider indistinguishable particles on a manifold $\mathcal{M}$. By definition of indistinguishability, our quantum state is fully specified if one knows that one particle is at position X_1 and another is at position X_2. The question of which particle is at X_1 and which is at X_2 does not make sense. So, the configuration space for N indistinguishable is $\mathcal{M}^N/S_N$, i.e., I have modded out the permutation group S_N.

When considering this many particle space it is natural to decompose the covariant derivative into single-particle and interaction parts. I use $\mathcal{A}$ to denote the many body $U(1)$ connection while A denotes the single particle $U(1)$ connection, i.e., the covariant derivative take the form

$$d + i\sum_{i=1}^{N} A(X_i) + i\mathcal{A}(X_1, \ldots, X_N) . \tag{2.6}$$

The form $A(X_i)$ is the same for all particles and does only depend on the position of the particle labeled by i, i.e.,

$$\begin{aligned} A(X_1) &= A_1(X_1)dx_1^1 + A_2(X_1)dx_1^2 \\ A(X_2) &= A_1(X_2)dx_2^1 + A_2(X_2)dx_2^2 \\ \vdots &= \vdots\,, \end{aligned} \tag{2.7}$$

where A_1 and A_2 have the same functional form in all rows. From now on, I assume that one can choose (and have chosen) $\mathcal{A}$ such that it is closed up to field strengths that vanish exponentially with $\min_{i,j}\left(|X_i - X_j|\right)/\lambda$, for some length λ. This assumption is not a restriction, since if no such choice is possible, there is no well-defined notion of topological interaction at all.

The topological interaction is defined by $\mathcal{A}$, and $\mathcal{A}$ is closed in the region where the distance between particles is much greater than λ, i.e., $\mathcal{A}$ is closed in $\left(\mathcal{M}^N - \delta\right)/S_N$, where

$$\delta = \left\{(X_1, X_2, \ldots)\,\middle|\, \min_{i,j}\left(|X_i - X_j|\right) \gg \lambda\right\}\,. \tag{2.8}$$

The gauge invariant information in $\mathcal{A}$ is captured by the phases $\left\{e^{i\int_\Gamma \mathcal{A}}\right\}_\Gamma$, where Γ is a closed curve. But since $\mathcal{A}$ is closed, the integral $e^{i\int_\Gamma \mathcal{A}}$ does not depend on continuous deformations of Γ. One thus need to consider the set of equivalence classes of closed curves in $\left(\mathcal{M}^N - \delta\right)/S_N$, where two curves are equivalent if there exists a continuous map which transforms one curve into the other.

There is one binary operation one can put on this set that is compatible with the mapping $\Gamma \to e^{i\int_\Gamma \mathcal{A}}$, and this operation is defined such that $\Gamma_1 \times \Gamma_2$ is the curve where you traverse first Γ_1 and then Γ_2. The compatibility of the operation and the mapping can easily be checked, since it is a direct consequence of the property of the line integral,

$$e^{i\int_{\Gamma_1\times\Gamma_2} \mathcal{A}} = e^{i\int_{\Gamma_1} \mathcal{A} + i\int_{\Gamma_2} \mathcal{A}} = e^{i\int_{\Gamma_1} \mathcal{A}} e^{i\int_{\Gamma_2} \mathcal{A}}\,. \tag{2.9}$$

So, the defining property $\left\{e^{i\int_\Gamma \mathcal{A}}\right\}_\Gamma$ of $\mathcal{A}$ is that it is a $U(1)$ representation of the group I just described.

The group that is the *fundamental group* of $\left(\mathcal{M}^N - \delta\right)/S_N$. The fundamental group of a manifold is illustrated in Fig. 2.2 and is denoted by Π_1, in this case $\Pi_1\left(\left(\mathcal{M}^N - \delta\right)/S_N\right)$.

Let us now be more specific: assume that $\mathcal{M}$ is some topologically trivial three-dimensional manifold and assume that there is N identical particles on $\mathcal{M}$. From Fig. 2.3 you should realize that any closed loop formed by a single particle can be contracted to identity. So all non-trivial loops must be built up by exchange of two particles, e.g., in the positive direction, i.e., *transpositions*. Figure 2.5 illustrates the fact that two transpositions can be continuously transformed into a loop where one

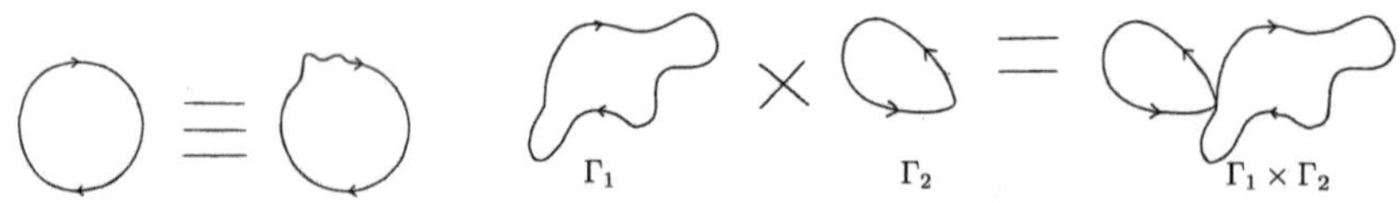

(a) The fundamental group of a connected manifold consist of equivalence classes of curves where two curves are equivalent if they can be continuously deformed into each other.

(b) The binary operator on the fundamental group maps the product of the equivalence class containing Γ_1 and the one containing Γ_2 to the one containing $\Gamma_1 \times \Gamma_2$, where $\Gamma_1 \times \Gamma_2$ is defined as the curve that first traverses Γ_1 and then Γ_2.

Fig. 2.2 The fundamental group

(a) **(b)** **(c)** **(d)**

Fig. 2.3 In three dimensions an encircling (i.e., the path in (**a**)) can continuously be contract to zero

particle encircle the other in the positive direction, i.e., *an encircling*. Thus, there is only one element in $\Pi_1\left(\left(\mathcal{M}^N - \delta\right)/S_N\right)$ for each permutation of the particles and I can therefore conclude that the fundamental group is S_N.

Let us see what the non-trivial $U(1)$ representations are. The permutation group is generated by transpositions, so it is enough to investigate what values you can get when integrating along a path Γ that results in a transposition. Since two transpositions equal identity you have $e^{i\int_\Gamma \mathcal{A}+i\int_\Gamma \mathcal{A}} = 1$, and one can conclude that there are two different possibilities: either you get a phase -1, or not, when you integrate $\mathcal{A}$ along a curve with an odd number of transpositions. These two possibilities correspond to fermions and bosons respectively.

Usually when dealing with fermions, one makes the analog thing as in (2.5), i.e., a singular gauge transformation, and consider the wave functions to be functions in the odd subclass of functions on $\mathcal{M}^N$ (i.e., the double cover of $\mathcal{M}^N/S_N$). One cannot consider the functions to be functions on $\mathcal{M}^N/S_N$, since with the singular gauge transformation there are two different function values for each point in $\mathcal{M}^N/S_N$—and these are related by a minus sign $e^{i\int_\Gamma \mathcal{A}}$, where Γ is a transposition.

One could also say, as is done in many introductory textbooks, that the wave functions for fermions are functions on $\mathcal{M}^N$ that have eigenvalue -1 with respect to the operator which interchanges two fermions. Although formally correct, this is confusing: it is nonsense to say that a fermion A is at position X and another fermion B is at position Y; they are indistinguishable and thus there is no physical way to, even in principle, tell which fermion is fermion A and which is fermion B. The operator which interchanges them thus cannot be a physical operator; it is only defined by the mathematical trick of using a double cover of the configuration space discussed in the previous section.

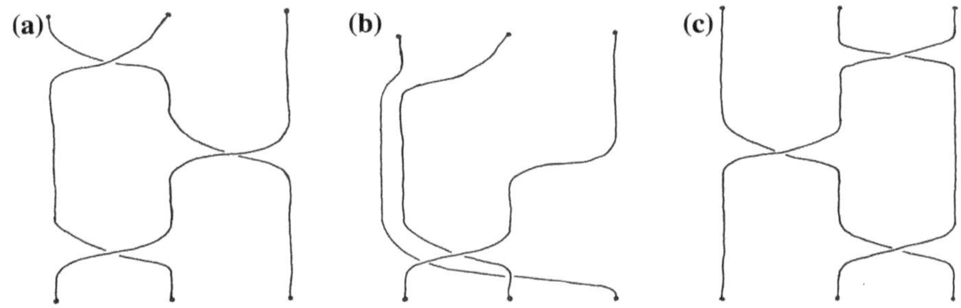

Fig. 2.4 In the figures above, each horizontal cross-section corresponds to a position on a loop. Figure **a** shows the loop corresponding to $T_i T_{i+1} T_i$. In figure **b** the loop from figure (**a**) was continuously transformed such that the particle who starts out to the right goes behind the other particles earlier. In figure **c** you should then recognize the loop as $T_{i+1} T_i T_{i+1}$, after another continuous transformation where the paths are straightend

I now assume that our base manifold $\mathcal{M}$ is a trivial, two-dimensional manifold. One can then no longer contract an encircling, and the fundamental group of $(\mathcal{M}^N - \delta)/S_N$ is not S_N, but a group called the *Artin Braid group*[2] or just the *braid group* (B_N).

From Fig. 2.5 you should realize that encirclings can be decomposed into two transpositions and, thus, the braid group of two particles, B_2, is just the free group generated by transpositions (T). With N particles you instead have N generators $\left(\left\{T_{(i,i+1)}\right\}\right)$, where $T_{(i,j)}$ corresponds to transposition of the particle labeled i and the one labeled j with no other particles in between.

If $|i - j| \geq 2$, transposing particle i and $i + 1$, and particle j and $j + 1$, is independent, and the order in which they are executed cannot matter. One thus have the relation

$$T_i T_j = T_j T_i \qquad ; \qquad |i - j| \geq 2 \,, \tag{2.10}$$

where I defined $T_i \equiv T_{(i,i+1)}$.

For $|i - j| < 2$ there is a less obvious relation:

$$T_i T_{i+1} T_i = T_{i+1} T_i T_{i+1} \,, \tag{2.11}$$

which can be understood from Fig. 2.4. So, the braid group is the free group generated by $\{T_i\}_i$ with the relations (2.10) and (2.11).

The braid group, as opposed to the permutation group, does not have only one non-trivial $U(1)$ representation. If one assumes that one only has one species ofanyons,

[2]After Emil Artin who discovered the braid group, see Ref. [12].

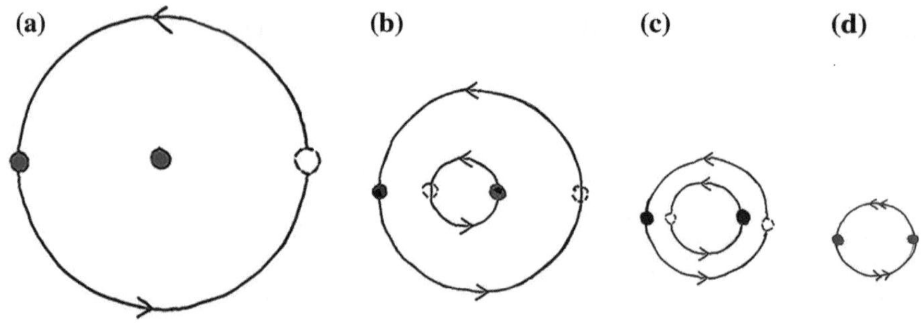

Fig. 2.5 In the above figures, the black dot denotes where the particle start, and the white dot denotes the half-way point on the path. Figure **a** illustrates an encircling which is continuously transformed into a double transposition

all permutations must be equivalent and the $U(1)$ representation $\left\{e^{i\int_\Gamma \mathcal{A}}\right\}$ is defined by the statistical angle θ,

$$R \equiv e^{i\theta} \equiv e^{i\int_{T(i,j)} \mathcal{A}} . \tag{2.12}$$

Generalizing this to the case with several different species of anyons is straightforward; for anyons of the same species there is no difference other than that you add a species label α, so that $R^{\alpha\alpha}$ corresponds to a transposition of two anyons of species α (Fig. 2.5).

One cannot define a transposition of anyons of different species, since transposing them would not return the same state. An encircling of anyons of different species is, however, well defined. I use $E_{(i,j)}$ to denote a curve corresponding to an encircling of an α-anyon labeled i and a β-anyon labeled j and the symber $R^{\alpha\beta}$ is defined as

$$\begin{aligned} R^{\alpha\beta} &\equiv e^{i\int_{T(i,j)} \mathcal{A}} \quad \text{if } \alpha = \beta \\ (R^{\alpha\beta})^2 &\equiv e^{i\int_{E(i,j)} \mathcal{A}} \quad \text{if } \alpha \neq \beta \end{aligned} \tag{2.13}$$

(Using a square in the definition of the encircling is just a practical convention.)

2.1.3 *Mediating Topological Interactions with a Gauge Boson*

Any two-body interaction can be rewritten as a mediation of an exchange-particle. Since both the braid group and the permutation group is generated by transpositions, which only involve two particles, the topological interaction is a two-body interaction, and $\mathcal{A}(X_1, \ldots, X_N)$ can be put in two-body form. I.e., there exists forms $\tilde{a}_{ij}$ such that

$$\mathcal{A}(X_1, \ldots, X_N) = \sum_{i \neq j} \tilde{a}_{ij}\left(X_i, X_j\right) . \tag{2.14}$$

Let us consider the system with one anyon species. It is then convenient to introduce $a(X_i)$, defined by

$$d_i a(X_i) = \frac{\theta}{2} \sum_{i \neq j} \mathcal{P}\left(X_j\right) , \tag{2.15}$$

where d_i is the exterior derivative acting only on the particle labeled i, and $\mathcal{P}$ i denote the Poincare dual (i.e., a delta-function form, see Definition 4 in Appendix B). Using this, the standard $K = p^2/2m$ kinetic term takes the form

$$K = \frac{1}{2m} \sum_{i=1}^{N} \left[d_i + iA\left(X_i\right) + ia\left(X_i\right)\right]^2 , \tag{2.16}$$

where the square is defined to be the usual Laplace-Beltrami operator defined by some spatial metric h, which also can be written as

$$\star_h \left(d - iA\left(X\right) - ia\left(X\right)\right) \star_h \left(d + iA\left(X\right) + ia\left(X\right)\right) \tag{2.17}$$

where I introduced the notationally convenient Hodge star operator $\star_h$ (see Definitions 8 and 10 in Appendix B). In second-quantized form this reads

$$\hat{K} = \frac{1}{2m} \int \psi^\dagger(X) \left(d - iA\left(X\right) - ia\left(X\right)\right) \star_h \left(d + iA\left(X\right) + ia\left(X\right)\right) \psi(X) , \tag{2.18}$$

which naively looks quadratic. But one has to remember (2.15), which in second quantized form reads

$$da = \frac{\theta}{2} \star_h \psi^\dagger \psi . \tag{2.19}$$

Solving for a and substituting back would give a quartic (i.e., interacting) term. Since it is quartic one can use a Hubbard-Stratonovich[3] transformation to decouple the interaction with an auxiliary field. With the form (2.18) the job is already done for us. The interaction is already in a decoupled form, one just have to introduce a Lagrange multiplier term that imposes the constraint (2.19).

So I introduce the Lagrange multiplier field a_t, and the result is[4]

---

[3] See e.g., page 197 of Ref. [13].

[4] I assume a notion of absolute rest and absolute time, implying that there exists a natural map relating spatial space-time forms to forms on a spatial slice and *vice versa*. (This is exactly what is done implicitly when the electric and magnetic vectors are extracted from the electromagnetic tensor). I will reserve the letter h for a spatial metric, and when it acts on space-time forms it should be understood that there is a composition with the map mentioned in the previous sentence. This notion is given in a more precise form in Definition 14, and the proceeding discussions, in Appendix B.

$$\mathcal{L} = \frac{1}{\theta} a_t dt \wedge da + a_t dt \wedge \star_h \psi^\dagger \psi \,. \tag{2.20}$$

This together with $\hat{K}$ in (2.18) is nothing but the Coulomb gauge version of the Chern-Simons Lagrangian

$$\mathcal{L} = \frac{1}{2\theta} a \wedge da + a \wedge \mathcal{J} \,, \tag{2.21}$$

where I redefined a to be the space-time version i.e.,

$$a = a_t dt + a_{x^1} dx^1 + a_{x^2} dx^2 \,, \tag{2.22}$$

and the current from $\mathcal{J}$ equals

$$\star_h \psi^\dagger \psi - \frac{i}{2m} dt \wedge \left(\psi^\dagger \star_h d\psi - \psi \star_h d\psi^\dagger\right) + \frac{1}{m} dt \wedge \psi^\dagger \star_h \left(A\left(X\right) + a\left(X\right)\right) \psi \,. \tag{2.23}$$

2.2 Topological Order in Two Dimensions

In this section I will discuss more general topological interactions and see why each topological interaction defines a distinct topologically ordered phase. As you will see a topological interactions is equivalent to an unitary representations of the braid group. This understanding was pioneered in Ref. [14] but brought to the physics community by Edward Witten, in Ref. [15], when he also introduced the Chern-Simons field theory (I will discuss the Abelian version later in this chapter).

I am not directly interested in classifying the topological interactions, but I will introduce the general mathematical structure that is used to specify a topological interaction, since it will teach us the most important ingredients and properties of topologically ordered states.

Since the topological information is independent of distance, I will, for simplicity, consider the special case where all particles are localized by some pinning potential and all dynamics is slow compared to $\hbar\Delta^{-1}$. Furthermore, I assume that the particles are well separated, that is, they are far enough away from each other to not interact (except for the topological interaction).

I will consider states

$$|X_1, \alpha_1; X_2, \alpha_2; \cdots\rangle \tag{2.24}$$

with the property that they are identical to the ground state $|GS\rangle$ at distances much further than the correlation length away from the points $\{X_i\}$. The difference from $|GS\rangle$ can be detected by a non-local operator, with support only far away (see Fig. 2.6).

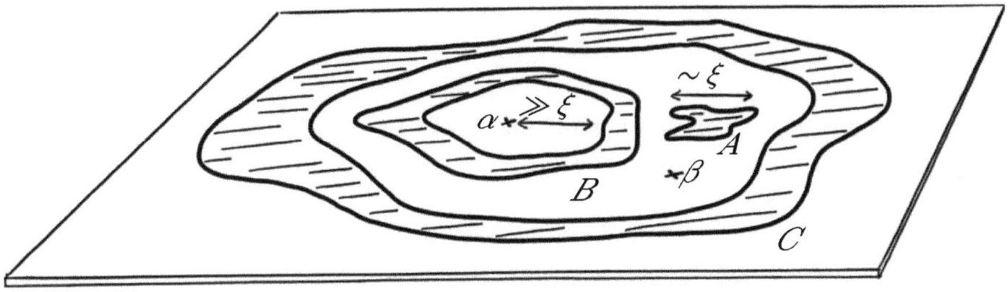

Fig. 2.6 Here α and β are anyons and ξ is the correlation length. **Regions of type** A, that is regions that only contain local information far away from any anyon, contain no information about the anyon types. **Regions of type** B, that encircle *one* anyon, do contain the information about the anyon type even if they are far away from the anyon. **Regions of type** C, that encircle several anyons, can contain more information than there is in all regions of type A and B. (By information in a region, say A, I mean everything that possibly could be deduced from the density matrix $\mathrm{Tr}_{X\notin A}|\psi\rangle\langle\psi|$.) *Figure by S. Holst*

In general, the label α_i denotes the type of anyon. If one has two states $|X;\alpha\rangle$ and $|X;\beta\rangle$ and they cannot be distinguished by an operator only acting far away then I define[5] $\alpha=\beta$. This means, e.g., that the deviation close to X_i (i.e., the particle) cannot be removed by any local operator $\mathcal{O}(X_i)$ only acting in the vicinity of X_i. For notational convenience, I also include the possibility that there is nothing at X_i that can, from far away, be distinguished from the ground state. In that case I write $\alpha_i=\mathbb{1}$.

I assume that all dynamics is generated by local Hamiltonians so that all particle states must be realizable by locally perturbing the ground state. Since the anyons are not local perturbations, it must be the case that they can be created in pairs (an *anyon, anti-anyon pair*) with the property that the pair is just a local perturbation when its constituents are close together.

I now consider time-evolution where anyons are moved around by being pinned to potentials and end up in the same configuration that they started in. I assume that the particles are far enough apart for all non-topological interaction (i.e., interaction which decrease with distance) to be negligible, and one can factorize the time-evolution operator U as

$$U=U_{top}\prod_i U_i\,, \tag{2.25}$$

where U_i are the single particle contribution to the time-evolution for particle i, and U_{top} is the topological time-evolution.

The topological interaction is quantized, and thus cannot change continuously. This means, that as long as one has a well defined adiabatic limit, i.e., the regime $k_BT\ll\Delta$, you cannot continuously interpolate between states with different

[5]Sometimes a different notation is used: if two states can be connected by some local operator $\mathcal{O}(\vec{x})$, $|\vec{x};\alpha\rangle=\mathcal{O}(\vec{x})|\vec{x};\beta\rangle$, then $\alpha=\beta$ and otherwise $\alpha\neq\beta$. The difference between this and the one in the text is that the one in the text treats a fermion just as a trivial anyon while this do not.

topological interactions. So, in the regime $k_B T \ll \Delta$, each topological interaction defines a specific state of matter—a topologically ordered state of matter.

Let us now see what specifies a topological interaction. If you consider a fixed number of particles, a topological interaction is by definition given by a unitary representation of the fundamental group

$$\Pi_1\left(\left(\mathcal{M}^{N_1+\cdots+N_n}-\delta\right)/S_{N_1}\cdots/S_{N_n}\right)\,, \tag{2.26}$$

which I will refer to as the n-species braid group. (I now included the possibility of several different types of indistinguishable particles.) The actual number of particles, $\{N_i\}_i$, is not a property of the phase, it is just a property of the particular state I consider at the moment. What I want is a compatible representation of all n-species braid groups with varying $\{N_i\}_i$.

The n-species braid group with particle numbers N_i is a subgroup of the n-species braid group with particle numbers M_i, if $N_i \leq M_i$. So finding the general unitary representation of varying number of particles is equivalent to finding it for $\Pi_1\left(\left(\mathcal{M}^{N_1+N_2+\cdots}-\delta\right)/S_{N_1}/S_{N_2}\cdots\right)$ with $N_i \to \infty$. I will refer to this group as the infinite n-species braid group, and a unitary representation thereof is specified by three tensors,

$$(N_\gamma^{\alpha\beta}, R_\gamma^{\alpha\beta}, F_{\delta\epsilon\zeta}^{\alpha\gamma\beta})\,, \tag{2.27}$$

a fusion tensor N, the braid rules R and the F-tensor. (Actually, N can be deduced from R and F, so one could do without it.)

Let us start by studying the fusion tensor. If two anyons fuse to a third anyon γ, I put $N_\gamma^{\alpha\beta} = 1$, and otherwise I put it to zero.[6] That the two anyons α and β fuse to γ means that when α and β are brought close together they cannot, from far away, be distinguished from γ.

Let us start by considering Abelian anyons. The only way to detect an anyon from far away is by the topological interaction and, as you saw in the previous section, the topological interaction among Abelian anyons is defined by the statistical angles (2.13). Thus, these statistical angles must also encode the fusion. Assume that you have exactly two non-trivial Abelian anyons, α and β, with $R^{\alpha\beta} = R^{\beta\beta} = (R^{\alpha\alpha})^2$. Then, there is no braiding operation that can distinguish a β from two α's very close to each other. Now we can conclude:

$$\alpha \times \alpha \equiv \alpha^2 = \beta\,, \tag{2.28}$$

or equivalently $N_\beta^{\alpha\alpha} = 1$. The fact that there is precisely two distinct non-trivial anyons also implies $R^{\alpha\alpha} = e^{i\pi/3}$, since otherwise α^3 would also be a non-trivial

[6]This definition is only valid for the special case when all $N_{\alpha\beta}^{\gamma}$ are either zero or one (all examples in this thesis will have this property). In general there are several locally indistinguishable states with three anyons α, β and γ that together fuse to identity, and in general the integer $N_{\alpha\beta}^{\gamma}$ is defined as the number of orthogonal such states.

anyon. With $R^{\alpha\alpha} = e^{i\pi/3}$ we, however, get $R^{\alpha^3\alpha} = R^{\alpha^3\beta} = \left(R^{\alpha^3\alpha^3}\right)^2 = 1$, so the fusion rules close on them selves:

$$\alpha \times \alpha \times \alpha = \alpha \times \beta = \mathbb{1}\,. \tag{2.29}$$

The above equation also tells us something more; that β is the anti-anyon, $\bar{\alpha}$, corresponding to α.

As you should realize from this example the fusion of Abelian anyons is trivial in the sense that it is specified directly by the statistical angles, so to continue one needs a non-Abelian example. I take the example of Ising topological interaction, since it will be relevant for our discussion of chiral p-wave superconductivity. With Ising topological interaction there are two anyon types σ and ψ, which in the p-wave SC context correspond to the fundamental vortex and the Bogoliobov particle (i.e., a broken Cooper pair) respectively. The fusion of α and β is not necessarily specified by the anyon types α and β. Or equivalently, the anyon types α and β of a state $|X, \alpha; Y, \beta\rangle$ specifies all properties of regions of type B (see Fig. 2.6), but not necessarily all properties of regions of type C.

This is always the case for non-Abelian anyons, and for Ising topological interaction the non-trivial fusions (i.e., the ones that are not zero and do not include fusion with the trivial anyon) is given by

$$\sigma \times \sigma = \mathbb{1} + \psi \qquad ; \qquad \sigma \times \psi = \sigma \qquad ; \quad \psi \times \psi = \mathbb{1}\,. \tag{2.30}$$

The anyon you would obtain if two anyons (α and β) are close together is called the *fusion channel* of α and β and the first rule mean that when you bring two σ's together there are two different fusion channels: they will either annihilate, or fuse to ψ.

Let us now consider the states with several anyons and resolve how many states that cannot be distinguished by sums of local operators (i.e., *locally indistinguishable states*).

In general, the fusion rules determine how the dimension of the space of locally indistinguishable states with a given number of anyons depends on the number and type of anyons. When pairing the σ's, two by two, there is, for each pair, a state where the constituents together form a ψ, and a state where they form a local excitation. These two different states are orthogonal, since they have different eigenvalues with respect to operators measuring the fusion channel of the σ's in the pair. Hence, the number of orthogonal and locally indistinguishable states corresponding to a given number of σ's grows two-fold each time two σ's are added.

If you bunch all σ's, two by two, such that each pair have a definite fusion channel, you have specified your state completely, since you then know a succession of unitary evolutions (the fusions to identity together with removing the local excitation) that bring us to the ground state. If you would have more complicated fusion rules, where the fusion products of the paired particles in turn have more than one fusion product, you would, to completely specify the state, also have to pair up the fusion products, and their products, *etc.*

Let us now understand the braid rules. Say that you have two σ's that are far away from all others. Transposing them cannot alter their fusion channel since that could be measured from far away, without getting close. The argument is general, thus the transposition, or encircling, of two particles in a distinct fusion channel cannot alter the state, and is therefore represented by a phase. For Ising anyons these phases are given by

$$R_1^{\sigma\sigma} = e^{-i\pi/8} \quad ; \quad R_\psi^{\sigma\sigma} = e^{i3\pi/8} \quad ; \quad R_1^{\psi\psi} = -1 \quad ; \quad R_\sigma^{\psi\sigma} = i\,, \tag{2.31}$$

where $U_{top} \doteq R_\beta^{\alpha\alpha}$ is the topological interaction corresponding to a transposition of two α anyons in fusion channel β and $U_{top} \doteq (R_\gamma^{\alpha\beta})^2$ corresponds to an encircling of an α around a β in the fusion channel γ. (The symbol $\doteq$ denotes "represented by".)

To know the representation of the infinite n-species braid group, you need to know how it acts in some basis. With a basis I here mean an orthogonal set of states that up to acting with local operators span the full low-energy Hilbert space. You can create a basis for any number of anyons by starting with a basis for a single anyon which is only labeled by anyon type, η. From this one-anyon basis you can create a basis of two anyons of type δ and ζ, specified by their type and their fusion channel η. This is done by acting with the operator $U_\eta^{\delta\zeta}$ that acts locally at η and splits η into fusion factors δ and ζ. Continuing, you can create a basis for three anyons, ε, γ and δ, with fusion channels $(\varepsilon, \gamma) \to \zeta$ and $(\zeta, \delta) \to \eta$ by acting with a local operator $U_\zeta^{\varepsilon\gamma}$ on ζ. Analogously continuing one step further by splitting ε into α and β you have a basis for four anyons specified by the types $(\alpha, \beta, \gamma, \delta)$ and the fusion channels $(\alpha, \beta) \to \varepsilon$, $(\varepsilon, \gamma) \to \zeta$ and $(\zeta, \delta) \to \eta$. This is depicted in the equation below where the "blob" in the notation is there to denote a region with an anyon that will split into fusion channels.

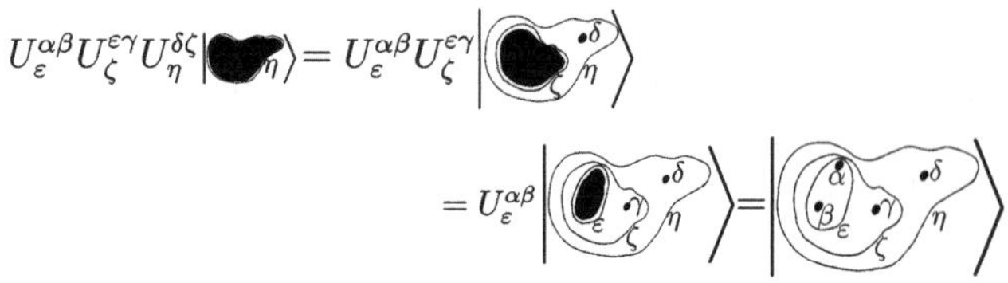

It should be clear how to continue this construction and get a basis for any number of anyons specified from the operators splitting anyons, $\{U_\gamma^{\alpha\beta}\}_{\alpha,\beta,\gamma}$, by starting with a basis for just one anyon. For specified anyon types $(\alpha, \beta, \gamma, \delta)$ you can use the above basis labeled by the fusion channels $(\varepsilon, \zeta, \eta)$:

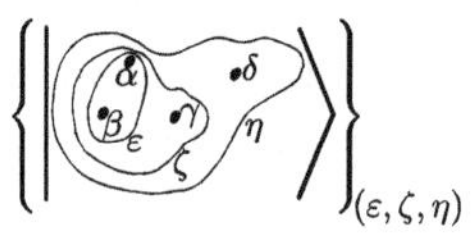

In this basis one knows the representation of, e.g., $U_{top}(T_{\alpha\beta})$; it is $U_{top}(T_{\alpha\beta}) \doteq R^{\alpha\beta}_{\varepsilon}$. But what is the representation of $U_{top}(T_{\beta\gamma})$? By a different set of operators acting on the one-anyon state, you can create another basis where $U_{top}(T_{\beta\gamma})$ is diagonal instead:

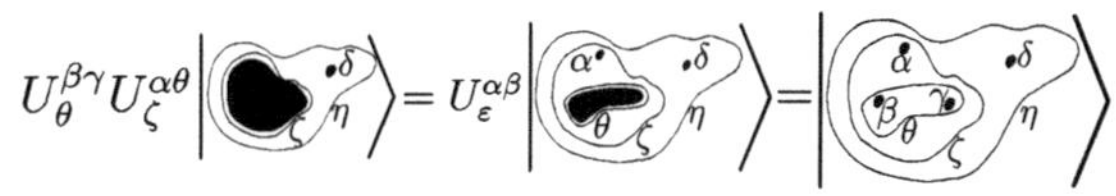

If you then know the overlap

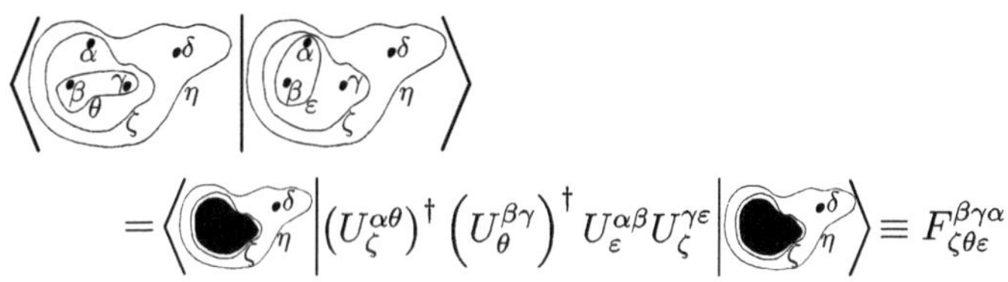

between the states in the different bases, we know how to transform between them and thus we know both the representation of $U_{top}(T_{\alpha\beta})$ and of $U_{top}(T_{\beta\gamma})$ in the same basis.

Since the information about the fusion channel ζ or the anyon type γ is not encoded locally at ε and $U^{\alpha\beta}_{\varepsilon}$ act locally on ε, its matrix elements cannot depend on neither ζ nor ε. The matrix elements of $U^{\alpha\beta}_{\varepsilon}$ can thus only depend on $(\alpha, \beta, \varepsilon)$ and similar for the other unitary operators in the equation above. That is why we can parametrize any such overlap with the tensor F, with only six indices.

So, by knowing the F-tensor we know how to change between bases where $U_{top}(T_{\beta\gamma})$ and $U_{top}(T_{\alpha\beta})$ respectively are diagonal, and we thus have a representation of both. With the same construction it is straight forward to use the F-tensor and the braid rules to get a representation for all braids for an arbitrary number of anyons, so the triplet (2.27) gives the full representation of the infinite n-species braid group.

2.2.1 *Torus Degeneracy*

Before we leave the general discussion concerning topologically ordered states in two dimensions, let us mention one more fact that dates back to work made already in the early nineties (see Refs. [16, 17]): on the torus, there are equally many locally indistinguishable ground states as there are numbers of anyon types. Here we include the trivial anyon as one of the anyon types—so, if there is one non-trivial anyon there is at least two locally indistinguishable ground states on the torus.

To see this, assume that we start out in a ground state $|GS\rangle$ and measure a topological property by locally creating an anyon, anti-anyon pair, dragging the anyon (β) around one of the non-trivial loops Γ of the torus (see Fig. 2.7), and then removing the anyon, anti-anyon pair with a local unitary time-evolution. The corresponding time-evolution operator has a well-defined topological part $U^{\beta}_{top}([\Gamma])$. (Here we stress that U_{top} does not depend on the specific curve Γ but on $[\Gamma]$, that is, the elements of the fundamental group).

By definition, $U^{\beta}_{top}([\Gamma])$ takes us from one ground state back to another.[7] We can with out loss of generality assume that the state we start out with is an eigenstate of $U^{\beta}_{top}([\Gamma])$,

$$U^{\beta}_{top}([\Gamma])|GS\rangle = \xi_1|GS\rangle \qquad ; \qquad \xi_1 \in \{x \in \mathbb{C}; |x| = 1\} \tag{2.32}$$

since if we did not return to the same state we could just diagonalize $U^{\beta}_{top}([\Gamma])$ in the finite Hilbert space

$$\text{span}\left[\left\{\left(U^{\beta}_{top}([\Gamma])\right)^n |GS\rangle\right\}_{n\in\mathbb{N}}\right] . \tag{2.33}$$

When we have done this we create an anyon (α), anti-anyon ($\bar{\alpha}$) pair. The previous described operation then comes in two varieties, $U^{\beta}_{top}([\Gamma])$ and $U^{\beta}_{top}([\tilde{\Gamma}])$: there are two different curves Γ and $\tilde{\Gamma}$ encircling the same direction on the torus, where $\tilde{\Gamma}$ can be deformed into Γ plus an encircling of α. We then define $U^{\beta}_{top}([\tilde{\Gamma}])|GS\rangle = \xi_2|GS\rangle$, and at least for some anyon β the phases are different, $\xi_1 \neq \xi_2$, since $U^{\beta}_{top}([\tilde{\Gamma}])$ and $U^{\beta}_{top}([\Gamma])$ is related by an encircling of α, which is detectable by braiding. When α then, is taken around the other circumference of the torus, see Fig. 2.7, more and more of the curves are in the equivalence class $[\tilde{\Gamma}]$. When it comes all the way around and is annihilated with $\bar{\alpha}$, there are no longer two different equivalence classes of

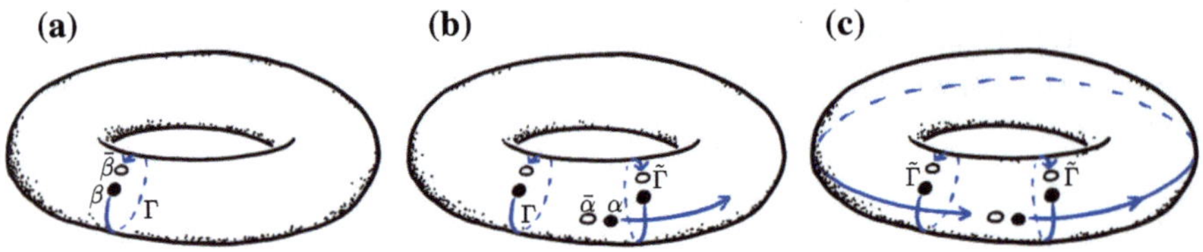

Fig. 2.7 Ground state degeneracy on the torus. **a** We measure a topological property of the ground state by creating an anyon (β) anti-anyon pair and dragging the anyon around a non-trivial loop Γ and annihilating it. **b** We create an anyon (α), anti-anyon ($\bar{\alpha}$) pair and then there is a different curve $\tilde{\Gamma}$ encircling the same direction on the torus, that can be deformed into Γ plus an encircling of α. **c** When α is taken around the torus and annihilated with $\bar{\alpha}$, there is no longer any curve corresponding to Γ, $\tilde{\Gamma}$ is everywhere, so to say. *Figure by S. Holst*

[7]There will always, because of finite-size effects, be a *true* ground state minimizing the energy of the *total* Hamiltonian, but we call all states that are locally indistinguishable from that state a ground state.

curves. And when now acting on the state with $U_{top}^{\beta}([\Gamma])$, we will no longer get the eigenvalue ξ_1, but ξ_2, and we can conclude that we are in a ground state orthogonal to the one we started with. If we continue this procedure we create as many states as there are numbers of anyon types.

From the procedure in which we got these state, we know that they are locally indistinguishable, so by definition the energy density in the states are the same. Since our Hamiltonian is local (i.e., exponential fall-off of the interaction as a function of distance), this also means that the total energy of these states has to be the same, up to exponentially small corrections in system size. This is one of the most important numerical signatures of topological order and something we will mention again later when discussing the chiral p-wave SC's in Chap. 4.

2.3 Abelian Chern-Simons Theory

All topological states we will consider are made up of real systems that at long length and time scales can be described by a field theory defined by some action. The action thus has to specify a specific representation of the infinite braid group, and in a coarse-grained field theory the topological interaction is all there is.

In this section we will consider a specific example of such a theory that we will come back to many times—the Abelian CS theory. On length scales much longer than the coherence length, and time scales much longer than $\hbar\Delta^{-1}$ (i.e., *the topological scaling limit*) the Abelian CS theory describes any system which only supports *Abelian* anyons. When studying this we will, among other things, be introduced to a very important concept, namely the bulk-boundary correspondence—i.e., we will see that the topological interaction in the bulk puts certain requirements on the edge theory.

The action is written in terms of N gauge fields $\mathbf{b} \equiv (b^1, b^2, \ldots)$ and an integer matrix $\mathbb{K}$ with non-zero determinant:

$$S = \frac{1}{4\pi} \int \mathbf{b}^T \wedge \mathbb{K} d\mathbf{b} \,. \tag{2.34}$$

Each gauge field corresponds to conservation of particle number of an anyon type. Breaking the gauge invariance means that there is no longer a conservation of particles. An anyon cannot be removed locally since it can be detected by measurements far away, so the topological interaction itself implies a conservation law. As we will see, the CS action implies a topological interaction, so the action has to be gauge invariant, otherwise it cannot be quantized to a unitary theory.

For the theory to be gauge invariant under large gauge transformations $b^I \to b^I + ie^{i\phi} de^{-i\phi}$, the matrix $\mathbb{K}$ must be integer-valued. This can be seen by an argument very similar to the discussion concerning large gauge transformations in the next chapter, (3.21). So, to avoid repetition the argument will be postponed until then. Furthermore, we can, without loss of generality, assume that the matrix $\mathbb{K}$ is symmetric, since if the term $K_{12}b_1 \wedge db_2$ is different from $K_{21}b_2 \wedge db_1$ we can just make a partial integration to make them equal.

Only gauge invariant quantities are physical and they are spanned by the Wilson loops,

$$W_{\Gamma,\mathbf{l}} = e^{i\int_\Gamma \mathbf{b}^T\mathbf{l}} , \tag{2.35}$$

with $\mathbf{l}$ being an integer vector and Γ a closed oriented curve (path ordering is omitted since the fields are Abelian so they commute anyway).

The action (2.34) does not contain any metric, or any other space-time structure, so the Wilson loops cannot depend on the precise path Γ but only on the topology of Γ. On this level there is no notion of which loops correspond to realizable physical processes or not—for that you would need to know more than just the topological theory. We should, however, still have a notion of space and time in mind; it is not as in, e.g., general relativity, where there is a real diffeomorphism invariance and no background metric at all.

Paths Γ that correspond to actual physical process, can be decomposed into paths which each can be parametrized by time. The number of paths needed depend on how many pair creation/annihilation processes we assume Γ correspond to. For simplicity we assume that Γ only correspond to one pair creation and one pair annihilation. Then it can be decomposed into two paths Γ_1 and Γ_2 that both can be parametrized by time. One path, Γ_1 , is directed forwards in time and one, Γ_2, is directed backwards. With that in mind we can write the Wilson loop as

$$W_{\Gamma,\mathbf{l}} = e^{i\int_\Gamma \mathbf{b}\cdot\mathbf{l}} = e^{i\int_{\Gamma_1} \mathbf{b}\cdot\mathbf{l}} e^{i\int_{\bar{\Gamma}_2} \mathbf{b}\cdot(-\mathbf{l})} , \tag{2.36}$$

where $\bar{\Gamma}_2$ denotes Γ_2 with flipped orientation. So, the interpretation of the Wilson loop is clear: it corresponds to creating a pair consisting of an anyon defined by the charge vector $\mathbf{l}$ and its anti-anyon (defined by $-\mathbf{l}$), and moving the anyon along Γ_1 and the anti-anyon along $\bar{\Gamma}_2$ until they meet and annihilate.

All topological interactions are generated by taking two different anyon, anti-anyon pairs, braiding the anyons once and annihilating them again. This is the statistical angle

$$e^{i\theta_{\mathbf{l},\mathbf{l}'}} \equiv e^{i2\pi\mathbf{l}\mathbb{K}^{-1}\mathbf{l}'} \propto \langle W_{\Gamma,\mathbf{l}} W_{\Gamma',\mathbf{l}'}\rangle , \tag{2.37}$$

where the curve Γ and Γ' are curves corresponding to the above process, i.e., the linking number between Γ and Γ' is one.

Since the theory is quadratic, all correlators can be calculated directly, so let us do that.

Since there is no path-ordering in the Wilson loops (2.35), we can write

$$\langle W_{\Gamma_1,\mathbf{l}_1} W_{\Gamma_2,\mathbf{l}_2} W_{\Gamma_3,\mathbf{l}_3} \cdots\rangle \propto \int D[\mathbf{b}]\, e^{iS[\mathbf{b}]+i\sum_i \mathbf{l}_i^T \mathbf{b}\wedge\mathcal{P}(\Gamma_i)} , \tag{2.38}$$

where $\mathcal{P}$ denotes Poincare dual (see Definition 4 in the Appendix B). Since Γ_i is closed we can write $\mathcal{P}(\Gamma_i) = -d\mathcal{P}(S_i)$, where S_i is a surface with Γ_i as its boundary. We can write the equations of motion as

$$\mathbf{b} = 2\pi \sum_i \mathbb{K}^{-1}\mathbf{l}_i\mathcal{P}(S_i) + d\lambda\,, \tag{2.39}$$

where λ is an arbitrary function. Plugging this expression for $\mathbf{b}$ into the action gives

$$\langle W_{\Gamma_1,\mathbf{l}_1} W_{\Gamma_2,\mathbf{l}_2} W_{\Gamma_3,\mathbf{l}_3} \cdots \rangle \propto \prod_{i<j} e^{i2\pi \mathbf{l}_i \mathbf{K}^{-1}\mathbf{l}_j \int \mathcal{P}(S_i)\wedge\mathcal{P}(\Gamma_j)}\,, \tag{2.40}$$

where the proportionality refers to single-particle effects, i.e., self-interactions that without regularization are not defined. The integral $\int \mathcal{P}(S_i) \wedge \mathcal{P}(\Gamma_j)$ gives the linking number (see Example 7 in Appendix B), which is half the number of transpositions of particle i and j, so the correlators are defined by how the particles have braided and the statistical angles.

The matrix $\mathbb{K}$ clearly specifies the theory, but does it provide only necessary information? If we do a change of functional variables

$$\mathbf{b} \to \mathbb{G}\mathbf{b} \qquad ; \qquad \mathbb{G} \in GL_N(\mathbb{Z})\,, \tag{2.41}$$

where $GL_N(\mathbb{Z})$ is the group of $N \times N$ integer matrices with integer matrix inverses, the matrix $\mathbb{K}$ changes as

$$\mathbb{K} \to \mathbb{G}\mathbb{K}\mathbb{G}^T\,. \tag{2.42}$$

So, we do not need to specify $\mathbb{K}$ completely, but rather only the equivalence class $[\mathbb{K}]$ under the equivalence reaction $\mathbb{K} \sim \mathbb{G}\mathbb{K}\mathbb{G}^T$, where $\mathbb{G} \in GL_N(\mathbb{Z})$. But this is still more than the necessary information. If we take a matrix $\mathbb{G} \in GL_M(\mathbb{Z})$, by definition $\mathbb{G}^{-1}$ is also an integer matrix and therefore

$$e^{i2\pi \mathbf{l}^T \mathbb{G}^{-1}\mathbf{l}'} = e^{i2\pi p} = 1 \qquad ; \qquad p \in \mathbb{Z}\,. \tag{2.43}$$

If we add M new fields $\tilde{b}^I$ and a term

$$\tilde{S} = \frac{1}{4\pi}\int \tilde{\mathbf{b}}^T \wedge \mathbb{G}d\tilde{\mathbf{b}}\,, \tag{2.44}$$

no statistical angles will be changed and it is thus the same topological order. Hence, we can conclude that the field theory is defined by matrices $\mathbb{K}$ under the equivalence relation that says $\mathbb{K} \sim \mathbb{L}$ if there exist matrices $\mathbb{G} \in GL_N(\mathbb{Z})$ and $\mathbb{H} \in GL_M(\mathbb{Z})$ such that

$$\mathbb{K} = \mathbb{G}\begin{pmatrix} \mathbb{L} & \mathbb{0} \\ \mathbb{0} & \mathbb{H} \end{pmatrix}\mathbb{G}^T\,. \tag{2.45}$$

As we said, the CS action cannot be quantized to a unitary theory without gauge invariance. But the theory is not invariant if put on a manifold $\mathcal{M}$ with a boundary $\partial\mathcal{M}$. If we perform a gauge transformation

$$\mathbf{b} \to \mathbf{b} + d\boldsymbol{\lambda}\,, \tag{2.46}$$

we get $S \to S + \delta S$ with

$$\delta S = \int_{\partial\mathcal{M}} \boldsymbol{\lambda}^T \mathbb{K} d\mathbf{b}\,. \tag{2.47}$$

One way out is to see what happens if one allows for non-gauge invariance on the edge. As opposed to the large gauge transformations one (almost) gets a well-defined unitary theory. One gets the chiral Luttinger liquid theory, see e.g., the references [18–20],

$$S_{edge} = \frac{1}{4\pi} \int dt\, dx\, \left[\partial_t \phi^T \mathbb{K} \partial_x \phi - \partial_x \phi^T \mathbb{V} \partial_x \phi\right]\,, \tag{2.48}$$

where the term defined by the matrix $\mathbf{V}$ is added as a regulator and is not universal—it is set by edge effects. The first term is however universal and is defined by the bulk up to the equivalence relation (2.45).

One can of course add different local terms on the edge which would change the edge physics. But there are some properties that one cannot get rid of. The situation is more complicated when $\mathbb{K}$ has mixed signs of its eigenvalues, so let us consider the simplest example, when the matrix $\mathbb{K}$ is chiral (e.g., all eigenvalues have the same sign). Then there is no way to gap out all the edge modes without breaking the $U(1)$ symmetries connected to the gauge invariance of the variables $\mathbf{b}$. If $\mathbb{K}$ has the minimal dimension to represent a certain topological order, then *all* gauge invariances are needed for the theory to be unitary and thus consistent. This is a first example of a bulk-edge correspondence; knowing the bulk topological order we can in some cases say with certainty that there will be, at least, a certain number of chiral edge modes!

References

1. J. Leinaas, J. Myrheim, Il Nuovo Cimento B Series 11 **37**, 1 (1977). https://doi.org/10.1007/BF02727953
2. D.C. Tsui, H.L. Stormer, A.C. Gossard, Phys. Rev. Lett. **48**, 1559 (1982). https://doi.org/10.1103/PhysRevLett.48.1559
3. B.I. Halperin, Phys. Rev. Lett. **52**, 1583 (1984). https://doi.org/10.1103/PhysRevLett.52.1583
4. D. Arovas, J.R. Schrieffer, F. Wilczek, Phys. Rev. Lett. **53**, 722 (1984). https://doi.org/10.1103/PhysRevLett.53.722
5. R.B. Laughlin, Phys. Rev. Lett. **50**, 1395 (1983). https://doi.org/10.1103/PhysRevLett.50.1395
6. G. Moore, N. Read, Nucl. Phys. B **360**, 362 (1991). https://doi.org/10.1016/0550-3213(91)90407-O
7. X.G. Wen, Phys. Rev. Lett. **66**, 802 (1991). https://doi.org/10.1103/PhysRevLett.66.802

8. M. Greiter, X. Wen, F. Wilczek, Nucl. Phys. B **374**, 567 (1992). https://doi.org/10.1016/0550-3213(92)90401-V
9. N. Read, D. Green, Phys. Rev. B **61**, 10267 (2000). https://doi.org/10.1103/PhysRevB.61.10267
10. D.A. Ivanov, Phys. Rev. Lett. **86**, 268 (2001). https://doi.org/10.1103/PhysRevLett.86.268
11. A. Stern, F. von Oppen, E. Mariani, Phys. Rev. B **70**, 205338 (2004). https://doi.org/10.1103/PhysRevB.70.205338
12. E. Artin, Abhandlungen aus dem Mathematischen Seminar der Universität Hamburg **4**, 47 (1925). https://doi.org/10.1007/BF02950718
13. A. Altland, B.D. Simons, *Condensed Matter Field Theory*, 2nd edn. (Cambridge University Press, Cambridge, 2010). https://doi.org/10.1017/CBO9780511789984
14. V. Jones, in *Geometric Methods in Operator Algebras*, ed. by H. Araki, E.G. Effros (Longman Scientific & Technical, Kyoto, 1983), p. 123
15. E. Witten, Commun. Math. Phys. **121**, 351 (1989)
16. X.G. Wen, Phys. Rev. B **40**, 7387 (1989). https://doi.org/10.1103/PhysRevB.40.7387
17. X.G. Wen, Q. Niu, Phys. Rev. B **41**, 9377 (1990). https://doi.org/10.1103/PhysRevB.41.9377
18. X.-G. Wen, Adv. Phys. **44**, 405 (1995). https://doi.org/10.1080/00018739500101566
19. X.-G. Wen, Int. J. Mod. Phys. B **06**, 1711 (1992). https://doi.org/10.1142/S0217979292000840
20. M. Stone, Ann. Phys. **207**, 38 (1991). https://doi.org/10.1016/0003-4916(91)90177-A

Chapter 3
Response Theory and Symmetry Protected Topological Phases

The topologically ordered phases, discussed in the last chapter, are the fundamental zero-temperature phases—if you allow for arbitrary changes of your system (i.e., the Hamiltonian), you can continuously interpolate between all other phases. But this is a too restrictive view and would make us miss important phase differences, as between solids and liquids. In many situations there are symmetries that *all* physically realizable perturbations, at least on long length scales, uphold. In those situations it is natural to consider what the possible phases are if we restrict ourselves to systems with a certain symmetry, i.e., *symmetry protected phases*.

The mere fact that we, if we break the symmetry, can continuously interpolate between the different symmetry protected phases means that there cannot, as in topological order, be an interplay between non-local and local degrees of freedom. Rather, there has to be a short-range entangled mechanism. We should say that we do not restrict our selves to studying the states with trivial topological order. We will also naturally encounter notable symmetry enriched phases namely fractional quantum Hall states.

The mechanism for symmetry protected phases will involve long length scales (i.e., much longer than the correlation length), but now we will not only consider topological interaction between particles; we will also look at the response to external structures.

We will start by understanding what symmetry protected phases are and focus on the perhaps most interesting (at least the most well studied) setting, namely $U(1)$ protected phases in two spatial dimensions. Their hallmark is the quantized Hall effect, and in the next section we will understand why the Hall conductance is quantized directly from microscopics.

Then we move on to a coarse-grained picture and use field-theoretic tools. There we use the notion of functional bosonization, which is a way to get a minimal theory which realizes a phase with a certain (in this case) $U(1)$ response in terms of a theory field which describes the conserved charge i.e., a *hydrodynamic theory*. This idea was coined in Ref. [1] which in turn was inspired by early hydrodynamic approaches, most notably Ref. [2].

T. Klein Kvorning, *Topological Quantum Matter*, Springer Theses,
https://doi.org/10.1007/978-3-319-96764-6_3

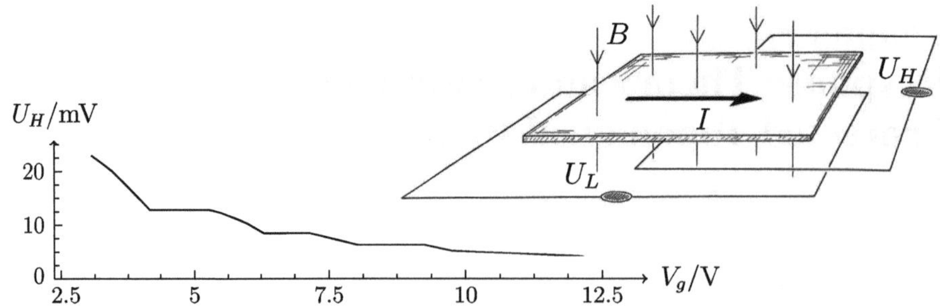

Fig. 3.1 Data from the original paper [3] with a set-up that schematically as depicted above. The chemical potential in the two dimensional electron gas increases monotonically with V_g and constant current is driven through the system. *Figure by S. Holst*

We will look at a nice application where one sees that the presence of topological order can solve an *apparent* gauge non-invariance (i.e., that the $U(1)$ charge is not conserved) which is present for fractional Hall conductances (what that is will become clear in the next section). This will also lead us to a previously unpublished result concerning how the field, which in functional bosonization usually is denoted by b, is normalized.

Finally, we move on to the interplay between the $U(1)$ symmetry and an external geometry. We will also generalize and no longer think of the $U(1)$ charge as the ordinary electromagnetic charge; in Sect. 3.3 the $U(1)$ charge under consideration will instead be electromagnetic flux. This sets the scene for the coming chapter.

3.1 The Quantization of the Hall Conductance

In this section we shall explain why the Hall conductance is quantized in $2d$ for gapped system at temperatures $k_B T \ll \Delta$. This is a most important fact: a quantized value cannot change continuously, so there has to be a phase transition between systems with different Hall conductances. Thus, the quantized value is a distinguishing feature of the different $2d$ $U(1)$ SPT phases.

Let us begin with a very famous experimental motivation. In Fig. 3.1 you can see a schematic experimental setting, as well as data from the original experiment performed by von Klitzing et. al. presented in Ref. [3]. The quantum Hall effect is realized in a two-dimensional electron gas, and the setup is configured such that the chemical potential of the gas increases monotonically with V_g (see Fig. 3.1). The experiment is performed at a temperature of 1.5 K and a magnetic field of 18 T. A constant current $I = 1\,\mu\text{A}$ is driven through the system and the perpendicular voltage U_H is measured. There are clear plateaus where the conductivity is constant, and on a closer inspection they can be seen to be integer multiples of the same constant, up to relative errors of the order $\lesssim 10^{-8}$. Now, let us make sense of this experiment, as well as get some useful theoretical insights that we will need later. The argument given here is bases on the work in Ref. [4] which for the integer quantum Hall effect gives the argument in the same generality as here, though phrased in a less modern language. (We here also inckude the fractional quantum Hall effect.) There are however many important earlier contributions that lay the ground work for the understanding, most notably the references [5–8].

We will assume that the Hall conductance is a well defined material property that only depend on macroscopic parameters, e.g., the density of impurities, the magnetic flux density *etc*. The Hall conductance is then independent of boundary conditions and we can assume that the geometry under consideration is a rectangle with periodic boundary conditions (i.e., a torus). We can also, in any way we like, change any parameters that locally cannot be detected, since by assumption this cannot alter the conductance. Note that even though the assumption made is powerful it is a very weak one; it only asserts that the materials under consideration have a well defined conductance in the first place.

With that said, we now consider a system which has an energy spectrum, on the torus, that has a gap $\Delta \gg k_B T$ above an N-dimensional subspace of degenerate states (up to splittings that vanish in the thermodynamic limit). We will also imagine having magnetic fluxes through the torus as illustrated in Fig. 3.2. Our previous assumption implies that for large systems, the conductance will not depend on these fluxes, and will also equal that for a physical Hall bar.

Let us begin by choosing a coordinate system for our torus, as depicted in Fig. 3.2. Inserting an arbitrary flux through any of the holes does not change the conductance, but it does change the Hamiltonian. However, for the special case of inserting a 2π flux, the resulting Hamiltonian is identical to the one where there is no flux. The Hamiltonian thus depends on the parameters $\phi_{x/y}$, which are defined on a space where the points (ϕ_x, ϕ_y) and $(\phi_x + 2\pi n_x, \phi_y + 2\pi n_y)$ are identified. Put differently, the parameter space $T_\phi^2 = \{(\phi_x, \phi_y)\}$ is a torus, which we will refer to as the *flux torus*, to distinguish it from the physical space which, because of the periodic boundary conditions, is also a torus.

The idea now is to consider mappings from the parameter space into the space of ground state wave functions. Such maps are characterized by an integer ch_1, called the first Chern number. The proper mathematical setting for this concept is the theory of *fibre bundles*. More precisely, the degenerate ground state wave functions form a complex vector bundle over the parameter space T_ϕ^2 (see Appendix C).

The basic result is that the Hall conductance σ_H, which we will define below, is given by the formula

$$\sigma_H = \frac{ch_1}{N}, \tag{3.1}$$

where N is the number of degenerate ground states. We are now ready for the actual calculation.

Let us first choose the gauge potential

$$A = \tilde{A} + \frac{\phi_x}{L_x} dx + \frac{\phi_y}{L_y} dy \tag{3.2}$$

where the integral of $\tilde{A}$ along any of the non-contractible loops on the torus is zero. Changing $\phi_{x/y} \to \phi_{x/y} + 2\pi$ would return the same physical Hamiltonian but in a different gauge, so $\phi_{x/y}$ is used not only to label the fluxes through the holes but also to set the gauge choice.

Now, assume that we have a monotonically increasing $\phi_y(t)$ such that in time τ, a full unit of flux is inserted in the hole, i.e., $\phi_y(\tau) = \phi_y(0) + 2\pi$. According to Faraday's law, this generates an electromotive force V_{Γ_y},

$$V_{\Gamma_y} = \frac{1}{2\pi} \int_{\Gamma_y} \mathcal{E} = \frac{\partial \phi_y}{\partial t}. \tag{3.3}$$

(For convenience, we have included a factor of 2π the definition of the electromotive force). Here, Γ_y is any of the non-contractible loops encircling the flux ϕ_y once, and $\mathcal{E}$ is the electric field one-form, (see Example 16 in Appendix B). The conductance is defined in the limit of vanishing electric field, so we should assume $\tau \to \infty$. Since there is an energy gap to all excited states, the time dependence of our quantum state is given by the adiabatic theorem, stating that the state will remain in an instantaneous ground state of the Hamiltonian. Let us now choose an orthonormal basis,

$$\left\{ |(\phi_x, \phi_y); \alpha\rangle \right\}_{\alpha=1,\ldots N} \Big|_{\phi_y=0}, \tag{3.4}$$

for the ground state manifold of the Hamiltonian at $\phi_y = \phi_y(0) = 0$, which is taken as a smooth function of ϕ_x. Under the adiabatic time evolution $U(t) \equiv U(\phi_y)$ (recall that $\phi_y(t)$ is monotonic) this evolves into,

$$|(\phi_x, \phi_y); \alpha\rangle = U(\phi_y) |(\phi_x, 0); \alpha\rangle. \tag{3.5}$$

Now we are ready to calculate the current. With no loss of generality we shall take the curve Γ_y to be a straight line in the x-direction and get

$$I_{\Gamma_y} = \langle\psi| \hbar^{-1} \left(\partial_{\phi_x} H\right) |\psi\rangle, \tag{3.6}$$

where we used the definition of the current density operator $j_x(X) = \partial H / \partial A_x = \partial H / \partial \phi_x$, and (3.2). We can without loss of generality assume that we start out in the state $|(\phi_x, \phi_y); 1\rangle$, which gives us

$$I_{\Gamma_y}(\phi_y) = \langle(\phi_x, \phi_y); 1| \hbar^{-1} \left(\partial_{\phi_x} H\right) |(\phi_x, \phi_y); 1\rangle. \tag{3.7}$$

Using the Schrdinger equation we know that acting with $i\hbar\frac{\partial\phi_y}{\partial t}\partial_{\phi_y}$ instead of the Hamiltonain yields the same result. This fact together with repeated use of Leibniz rule and the definition $E_0 = \langle(\phi_x, \phi_y); 1|H|(\phi_x, \phi_y); 1\rangle$ results in

$$I_{\Gamma_y}(\phi_y) = \hbar^{-1}\partial_{\phi_x}E_0 + i\frac{\partial\phi_y}{\partial t}\varepsilon^{ij}\partial_{\phi_i}\langle(\phi_x, \phi_y); 1|\partial_{\phi_j}|(\phi_x, \phi_y); 1\rangle. \tag{3.8}$$

If we integrate over the fluxtorus to get the average current, the first term vanishes since $E_0(\phi_x, \phi_y) = E_0(\phi_x + 2\pi, \phi_y)$. Up to the factor $\frac{\partial\phi_y}{\partial t}$, the second term equals a term in the trace of the Berry field strength corresponding to the state $|(\phi_x, \phi_y); 1\rangle$ (see Eq. (C.16) in Appendix C).

Now, we argue that the result is the same if we take the average after we have repeated the above calculation with all (i.e., $\alpha = 1, 2, \dots$) of the ground states as initial states. To see this let us to begin with assume that there are two ground states, and compare their conductivity in some bounded region. (This can in principle can be measured by a local probe.) One possibility is that the conductivity is in fact the same in the two states, and then the conductance trivially equals the average of the conductances of the two states. The other possibility is that there is a region where the conductivities do differ, which means that there are *local* operators with different expectation values in the two states. Now think of adding such terms to the Hamiltonian in some region. This will break the degeneracy, and result in a unique ground state. If we now calculate the conductance using a loop Γ far away from the region where the Hamiltonian was changed, we will again arrive at (3.1), for each one of the states separately, but with $N = 1$. In this case, the total conductivity will just be the sum of the two contributions. In both cases we can thus replace the second term in (3.8) with the average $\mathrm{Tr}(\mathcal{F})/N$, to get

$$\bar{I}_{\Gamma_y} = \frac{1}{2\pi\tau}\int_0^{2\pi}\int_0^{\tau} d\phi_x dt\, I_{\Gamma_y}(\phi_y, t) = \frac{1}{\tau N}\int_{T^2_\phi}\frac{\mathrm{Tr}(\mathcal{F})}{2\pi} = \frac{1}{\tau N}ch_1. \tag{3.9}$$

We get the average electromotive force, $\bar{V}_{\Gamma_y}$, directly from (3.3) and it equals τ^{-1}, so we finally get,

$$\sigma_H = \frac{\bar{I}_x}{\bar{V}_{\Gamma_y}} = \frac{ch_1}{N}, \tag{3.10}$$

which concludes the proof of (3.1).

Note that this proof only relied on the conservation of charge and that we were in a ground state of a gapped[1] Hamiltonian, i.e., $\Delta \gg k_B T$. Since the ground state degeneracy cannot vary without leaving this physical domain, each value of the conductivity corresponds to a phase of matter! These phases of matter are exactly

[1] A gap to all states is actually not needed. To be more precise, we need a gap to all exitations that can transport current. In a real quantum Hall system there is generically both charged excitations that are local and gapless excitations that cannot transport charge.

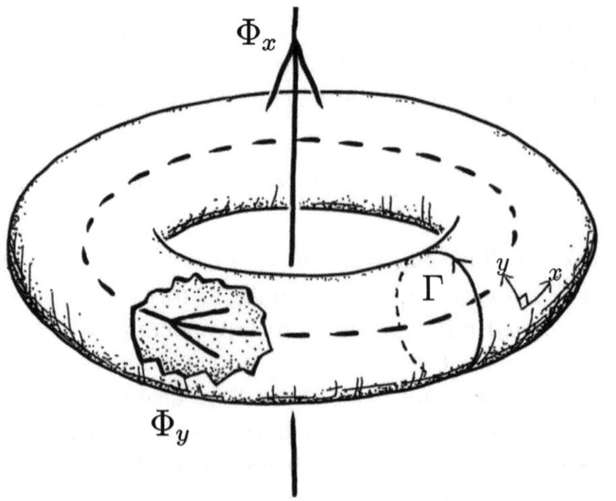

Fig. 3.2 Flux through the holes of the torus and an example of a curve Γ which encircle the flux Φ_y. *Figure by S. Holst*

the symmetry protected topological phases we have talked about. We will now move on and see how we can repruduce this using a coarse-grained picture.

3.2 $U(1)$ Response Theory

Our starting point is the partition functional, which in path integral notation can be written as

$$\mathcal{Z}[A^{ext}] = \int \mathcal{D}[\phi] e^{iS[\phi, A^{ext}]}. \tag{3.11}$$

It encodes all information about current correlation functions in the presence of a background field A^{ext}: the connected correlation functions are given by

$$\langle \mathcal{J}(X_1)\mathcal{J}(X_2)\cdots\rangle_{con} = \frac{\delta}{\delta A^{ext}(X_1)} \frac{\delta}{\delta A^{ext}(X_2)} \mathcal{W}[A^{ext}_\mu], \tag{3.12}$$

where $\mathcal{W}[A^{ext}] = -i \log \mathcal{Z}[A^{ext}]$ and $\frac{\delta}{\delta A^{ext}(X_1)}$ is the functional derivative, see Definition 3 in Appendix B. We are not interested in the exact microscopic currents but in coarse-grained quantities, e.g., $\bar{\rho}(X) = \frac{1}{Vol(\mathcal{N}_X)} \int_{\mathcal{N}_X} d^2x\, \rho(X)$ where $\mathcal{N}_X$ is some neighbourhood around X. This is equivalent to keeping only the terms with the lowest number of derivatives in the functional $\mathcal{W}[A^{ext}]$.

At the same time, we are only interested in the response for small changes away from some fixed background $A^{b.g.}$, meaning we can keep only the terms of the lowest powers in the fields. With that said, let us first consider the most important term for trivial insulators, i.e., the one with lowest number of derivatives and powers of the external fields. In ordinary insulators, the lowest order terms are quadratic in the electric and the magnetic fields. These terms, which we will refer to as *Maxwell terms,* are the terms with the lowest number of derivatives, which are quadratic in the fields (and thus give linear response), and is invariant under rotations, reflections (parity) and time reversal. Thus, they describe the response of a large class of isotropic materials to weak and slowly varying electromagnetic fields. If we, however, have a chiral system in $2d$, there is one more term relevant at even longer length scales—the Chern-Simons term, which encodes the Hall conductivity. The Hall conductivity σ_H is defined by the relation

$$\mathcal{J}^{sp} = \frac{\sigma_H}{2\pi}\mathcal{E}, \tag{3.13}$$

where $\mathcal{J}^{sp}$ is the spatial current form, see the Definition 15 in Appendix B. Combining this with the source-free Maxwell law, $0 = dF = d\mathcal{B} + dt \wedge d\mathcal{E}$, and current conservation $0 = d\mathcal{J} = d\varrho + dt \wedge d\mathcal{J}^{sp}$, we get

$$\mathcal{J} = \frac{\sigma_H}{2\pi}F\,. \tag{3.14}$$

(This is assuming that we start out with zero charge at time $t \to -\infty$.)

Written in this way, it should be obviouse why the Hall conductivity is special. It contains no metric information—it just enforces a proportionality between two two-forms. Without any other geometric structure, this is the only relation we can have between the current form and the electromagnetic field tensor. If we assume that averages on long enough scales are independent of the metric, this relation has to hold, or to be more precise

$$\int_S \mathcal{J} = \frac{\sigma_H}{2\pi}\int_S F \tag{3.15}$$

for large enough surfaces S.

Note that if σ_H is a constant, current conservation is ensured directly by Maxwells source-free law. But, if it is not, we do not have current conservation for a general external field, so from current conservation we can conclude that σ_H has to be a constant for (3.14) to be consistent. It can thus neither vary spatially, nor can it vary temporally by adiabatically changing our system.

By integration of the relation (3.14), we get the Chern-Simons response term:

$$\mathcal{W}_{CS}[A^{ext}] = \frac{\sigma_H}{4\pi}\int A^{ext} \wedge dA^{ext}\,. \tag{3.16}$$

As opposed to the Maxwell terms, this functional violates both time reversal, and parity symmetry. We thus conclude that in a system where these symmetries are present, the Hall conductance is zero.

What is even more important to is that (3.16) is not written in terms of field strengths only, so one might worry that it is not gauge invariant, which would imply that the $U(1)$ current is not conserved. Under the gauge transformation

$$A^{ext} \to A^{ext} + d\lambda \tag{3.17}$$

we get the variation

$$\delta \mathcal{W}_{CS} = \frac{\sigma_H}{2\pi} \int_{\mathcal{M}} d\lambda \wedge dA^{ext} = \frac{\sigma_H}{2\pi} \int_{\partial\mathcal{M}} \lambda dA^{ext}, \tag{3.18}$$

where $\partial\mathcal{M}$ is the boundary of the pertinent space-time $\mathcal{M}$. So, if we have a system that reaches out to infinity we can think of it as a three-sphere an there is no problem with gauge invariance (below we will discuss what happens at the boundary).

The important thing to notice here is that this gauge invariance does only hold since the parameter σ_H is constant both in space and time, which means that for a gapped energy spectrum it cannot vary across the system, neither can it vary with time as we adiabatically change any microscopic parameters defining our system. The only way to change it is to either close the gap or break the $U(1)$ symmetry. So, by definition, each value of σ_H corresponds to a specific $U(1)$ protected phase of matter.

In case of a boundary, we no longer have gauge invariance. This means that a system with the properties that we described, i.e., a gap to charge excitations and a non-zero Hall conductance, cannot have a boundary. The outcome is that if you have a system that on closed surfaces are gapped and have the Chern-Simons response, then there will automatically be gapless anomalous edge modes with an anomaly that will precisely cancel the gauge invariance from the Chern-Simons term. This could actually be taken as the defining feature for a symmetry protected topological phase i.e., that there is some anomalues response in the bulk that only can be consistent by the addition of an also anomalous edge, see Ref. [9].

Even without edges, there is also another gauge non-invariance present if σ_H is not an integer. To see this, let us assume that we have boundary conditions such that we start at $t = -\infty$ with zero external field, and also end up at $t = +\infty$ with zero external field. In that case we can view the time direction as being closed (not assuming this would produce the same kind of non-gauge invariance as if we had open boundary conditions). We assume the spatial space to be a torus and let

$$A^{ext} = \frac{\phi(t)}{L_y} dy, \tag{3.19}$$

where y is one of the circumference directions on the torus and L_y is the circumference in that direction; $\phi(-\infty) = 0$ and $\phi(+\infty) = 2\pi$. This is compatible with our boundary conditions: remember that, by the gauge transformation $A^{ext} \to A^{ext} + ie^{2\pi i y/L_y} de^{-2\pi i y/L_y}$,

$$A^{ext} = \frac{2\pi}{L_y} dy \tag{3.20}$$

is equivalent to $A^{ext} \equiv 0$. Let us now assume that field configuration and see what happens to the partition functional when we do the gauge transformation

$$A \to A + ie^{2\pi i x/L_x} de^{-2\pi i x/L_x}. \tag{3.21}$$

We get

$$\int A^{ext} \wedge dA^{ext} \to \int A^{ext} \wedge dA^{ext} + (2\pi)^2 \tag{3.22}$$

which corresponds to the change

$$\mathcal{Z}[A^{ext}] \to \mathcal{Z}[A^{ext}]e^{2\pi i \sigma_H}, \tag{3.23}$$

of the partition functional $\mathcal{Z} = \exp i\mathcal{W}_{CS}$. We can thus conclude that putting our system on a torus is not compatible with σ_H not being an integer. The solution to the problem is the assumption that there is a gap to a unique ground state. We will see that if we put our system on a torus, there has to be degenerate ground states.

3.2.1 *Functional Bosonization*

We have seen that the Chern-Simons term tells us things about what will happen to our system when we put the system on different geometries. Is there a way to go backwards and get a model system which is compatible with having a Chern-Simons response?

The starting point to answer this is to consider the gauge invariance of the response functional. The response functional is invariant under gauge transformations, so we have the equality

$$\mathcal{Z}[A^{ext}] = \mathcal{Z}[A^{ext} + a], \tag{3.24}$$

if a is restricted such that all Wilson loops $e^{i\int_\Gamma a}$ are unity. So, we can write

$$\mathcal{Z}[A^{ext}] \propto \int \mathcal{D}[a]_{res.} \mathcal{Z}[A^{ext} + a], \tag{3.25}$$

where the integral is taken over such restricted one-forms a. The idea is now to insert a delta functional, which ensures this restriction, and let the integral over a be free. Let us see if the functional

$$\int \mathcal{D}[F_b]_c e^{\frac{i}{2\pi}\int a \wedge F_b} \tag{3.26}$$

does the trick. Here, the subscript c denote that the integral runs over all closed forms F_b. The factor $1/2\pi$ is for convenience, as will become clear in a moment. Since the integral is over all closed forms, it also runs over all exact forms $F_b = db$, so let us first see what the integral would give if the domain was solely over the exact forms. We would have

$$\int \mathcal{D}[b] e^{\frac{i}{2\pi}\int a\wedge db} = \int \mathcal{D}[b] e^{\frac{i}{2\pi}\int da\wedge b} = \delta(da), \tag{3.27}$$

which would be enough to ensure that all Wilson loops are zero (if it where not for non-contractible loops).

Let us now consider a different restriction of the domain of the integral (3.26), namely when $F_b = \lambda\mathcal{P}(\Gamma)$, where Γ is some closed non-contractible loop and λ is a real number. Then we would get

$$\int d\lambda e^{\frac{i\lambda}{2\pi}\int a\wedge\mathcal{P}(\Gamma)} = \int d\lambda e^{\frac{i\lambda}{2\pi}\int_\Gamma a} \propto \delta\left(\int_\Gamma a\right). \tag{3.28}$$

But this is too restrictive! The requirement should be that the Wilson loops, $e^{i\int_\Gamma a}$, equals unity, and we should have the delta function

$$\sum_n \delta\left(\int_\Gamma a - 2\pi n\right). \tag{3.29}$$

This is what we would get if the integral over λ were replaced with a sum over the domain $\lambda \in 2\pi\mathbb{Z}$, which is implied by

$$\int_S F_b = 2\pi n. \tag{3.30}$$

This is just the Dirac quantization condition, i.e., that $\int_S F_b$ could be interpreted as a Chern number for a $U(1)$ connection. So, the integral we are after is

$$\int \mathcal{D}[b] e^{\frac{i}{2\pi}\int a\wedge db}, \tag{3.31}$$

where b should be thought of as a $U(1)$ connection.

Note that if we would have had a different coefficient than $1/2\pi$, in (3.26), then we would have gotten a different condition in (3.30) and thus not the convention for normalization of $U(1)$ connections used in this thesis. So, with this interlude we realize that the partition function can be written on the form

$$\mathcal{Z}[A^{ext}] = \int \mathcal{D}[a]\mathcal{D}[b]\mathcal{Z}[A^{ext} + a] e^{\frac{i}{2\pi}\int a\wedge db}, \tag{3.32}$$

where b is a $U(1)$ connection. Now we shift the integration variable, $a \rightarrow a - A^{ext}$, and assume that $\mathcal{Z}[A^{ext}] = e^{i\mathcal{W}_{CS}[A^{ext}]}$. If we then integrate out a, we get

$$\mathcal{Z}[A^{ext}] = \int \mathcal{D}[b] e^{\frac{i}{4\pi\sigma_H}\int b\wedge db + A^{ext}\wedge db}. \tag{3.33}$$

As we said, the Chern-Simons action is only valid at on a boundary-less manifold that has trivial fundamental group. Hence, there is no reason that the above procedure should work in a more general setting. But if we remember our discussion concerning CS theory from the last chapter, we realize that the partition functional (3.33) is for an arbitrary geometry a well-defined functional integral in the variables b, if we take $\sigma_H^{-1} = m \in \mathbb{Z}$. So we have managed to get a hydrodynamic theory which have the correct $U(1)$ response.

With this excursion into the relation between Chern-Simons theory and Hall conductance, let us consider the more general case of a Chern-Simons theory with some preserved $U(1)$ charge. With a metric, we can write down many terms that couple gauge fields b to the external gauge field A, but if we are interested in the coarse-grained response of very smooth fields, we are left with only the minimal coupling defined by a vector $\mathbf{t}$:

$$S[A^{ext}] = \frac{1}{4\pi}\int \mathbf{b}^T \wedge \mathbb{K} d\mathbf{b} + \frac{1}{2\pi} A^{ext} \wedge \mathbf{t}^T d\mathbf{b}. \tag{3.34}$$

Using similar arguments as before we realize that the partition function

$$\mathcal{Z}[A^{ext}] = \int D\,[\mathbf{b}]\, e^{iS[A^{ext}]}, \tag{3.35}$$

will only be gauge invariant under large gauge transformations if $\mathbf{t}$ is an integer vector. We can integrate out the $\mathbf{b}$ fields in the same way as we did in the last chapter, and then identify

$$\sigma_H = \mathbf{t}^T \mathbb{K}^{-1} \mathbf{t}. \tag{3.36}$$

We can also get the charge of the excitations. If we take a particle around a loop Γ, that do not encircle any anyon, the Berry phase divided by the electromagnetic flux through the loop defines the charge:

$$Q_{\mathbf{l}} = \mathbf{t}^T \mathbb{K}^{-1} \mathbf{l}. \tag{3.37}$$

Here, $Q_{\mathbf{l}}$ is the charge of a particle of type $\mathbf{l}$.

3.3 The Wen-Zee Term and Chiral Superconductors

There is always at least one more long-range structure apart from the external $U(1)$ field, namely geometry. Let us study how the partition function depends on the geometry of our surface in question:

$$\mathcal{Z}\left[A^{ext}, h\right] = \exp\left(i\mathcal{W}[A^{ext}_{\mu}, h]\right). \tag{3.38}$$

Here, h denotes the spatial (possibly time dependent) metric.

It has been shown in Ref. [10] that the coarse-grained version of this response is the so-called Wen-Zee term (discoverd in Ref. [11])

$$\mathcal{W}_{WZ}[A^{ext}, \omega] = \frac{\kappa}{2\pi}\int \omega \wedge dA^{ext}, \tag{3.39}$$

where ω is a potential for the Gauss curvature K, $d\omega = \star_h K$ (and thus is defined by h). When discussing superconductors, which are flux insulators, we will refer to this response as *the geometric Messiner effect*, see Ref. [12]. Since then this term will assert that there will be a spontanouse magnetic field in the presence of curvature, see Fig. 3.3. But now when we discuss charge insulators it gives rise to a shift in the relation between the total charge and the total flux,

$$N_Q = \nu N_\phi + \kappa\chi; \qquad \chi \equiv \frac{1}{2\pi}\int_S d\omega. \tag{3.40}$$

Note that, just as the Chern-Simons term, the Wen-Zee term specifies a specific orientation given by the sign of κ, so it can only be present if there is a preferred orientation—given by, e.g., an external magnetic field or by spontaneous symmetry breaking.

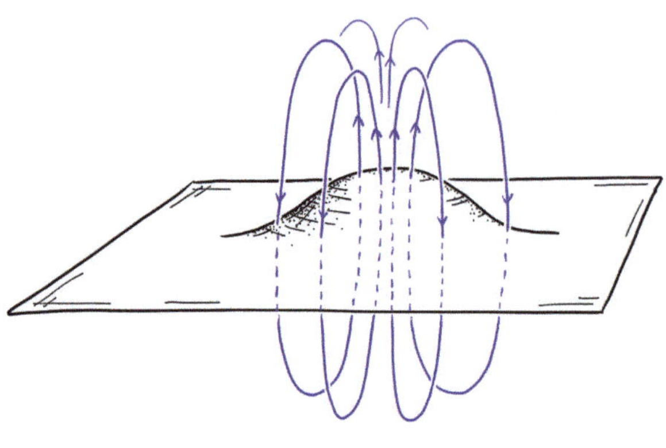

Fig. 3.3 An illustration of the geometric Meissner effect. This is the relevant 3d picture if the London length is short compared thickness of the sheet. *By S. Holst*

This physics can, again, be captured by a minimal coupling of a Chern-Simons theory to the spin connection,

$$S[A^{ext}, h] = \frac{1}{4\pi} \int \mathbf{b}^T \wedge \mathbf{K} d\mathbf{b} + \frac{1}{2\pi} A^{ext} \wedge \mathbf{t}^T d\mathbf{b} + \frac{1}{2\pi} \omega \wedge \mathbf{s}^T d\mathbf{b}. \tag{3.41}$$

Integrating out the **b**-fields we can read of the κ parameter from equation on top of the page,

$$\kappa = \mathbf{s}^T \mathbf{K}^{-1} \mathbf{s}. \tag{3.42}$$

If we take a particle around a loop Γ, that do not encircle any other anyons or enclose any electromagnetic flux, the Berry phase generated, divided by the total curvature through the loop defines the spin,

$$S_{\mathbf{l}} = \mathbf{s}^T \mathbb{K}^{-1} \mathbf{l}. \tag{3.43}$$

Here $S_{\mathbf{l}}$ is the spin of a particle of type **l**.

3.3.1 Chiral Superconductors

Now, when we have revised the topological response of quantum Hall insulators, let us switch to the system of interest.

Maxwell's source-free equation implies conservation of magnetic flux, which in two spatial dimensions is nothing but an ordinary $U(1)$ symmetry,

$$\mathcal{J}^{flux} \equiv \frac{F}{2\pi}. \tag{3.44}$$

Since SC's have a gap to flux excitations, one can consider them as flux insulators. We can thus consider the response to an external gauge field b^{ext} coupling to $\mathcal{J}^{flux}$. From the definition of $\mathcal{J}^{flux}$ we can conclude that db^{ext} is nothing but an external charge current $2\pi\mathcal{J}^{ext}$. Let us now discuss the different terms above, which are well known in the quantum Hall context but so far not in this SC context. We consider

$$\mathcal{W}_{CS}[b^{ext}] = \frac{\sigma_H^{flux}}{4\pi} \int b^{ext} \wedge db^{ext} \tag{3.45}$$

and

$$\mathcal{W}_{WZ}[b^{ext}, h] = \frac{\kappa^{flux}}{2\pi} \int \omega \wedge db^{ext}. \tag{3.46}$$

As said above, both these terms imply an orientation and could thus only be present in chiral SC's. As before, each coupling coefficient corresponds to chiral superconducting phases. The flux Hall effect would be much more difficult to measure than the ordinary charge Hall effect, but it would still be theoretically interesting if it was present. However, it seems unlikely, though, that there is a system which has non-trivial flux Hall coefficient.

We will instead focus on the Wen-Zee term, which in the quantum Hall context mostly is of pure theoretical interest, but could actually be measured directly and should generically be present in chiral SC's. The reason is that the analog of the Eq. (3.40) is:

$$N_\phi = \frac{\kappa}{2}\chi, \tag{3.47}$$

where again N_ϕ is magnetic flux (measured in units of 2π) and χ is the Euler charactersitic, i.e., the total curvature measured in units of 2π. This means, that through any curved region of a surface, there will be a spontaneous magnetic flux proportional to the total curvature of that region. This is a shift from zero, which means that it should be very easy to distinguish *having* this property from *not having* this property!

Already the notion of a chiral SC maybe seems like a quite exotic phase and this analysis is made for pure $2\mathscr{d}$ systems with $2+1\mathcal{D}$ electromagnetism. Because of this you might think that the analysis is purely of theoretical interest; just to discuss what phases that in principle are possible. But let is at least have a look at the experimental situation.

For the presence of a geometric Meissner effect it does not matter if the pairing is odd (as e.g., p-wave) or even, (e.g., d-wave) but it does make a big difference when it comes to the topological order which we will discuss in the next chapter. (The odd chiral paired states support non-Abelian anyons while the even ones support Abelian anyons, see References [13, 14].) Evidence for odd chiral pairing has been seen in many materials such as UPt_3 (see Ref. [15]) Li_2Pt_3B (see Ref. [16]), *etc.* The most well studied is however Sr_2RuO_4 (see Ref. [17] for a summary of experiments done one Sr_2RuO_4). There are also many candidate materials with even chiral pairing such as SrPtAs (see references [18, 19]), doped graphene (see references [20–22]), $Na_xCoO\cdot yH_2O$ [23] etc.

In all those materials electromagnetism is of course $3\mathscr{d}$ and even though there are ideas how make electromagnetism $2\mathscr{d}$ (see e.g., [24]) it is not needed. Considering the case of a $3\mathscr{d}$ SC, we still have a gap to flux excitations, and flux will still be conserved, although it is string-like rather than point-like. And, if the $3\mathscr{d}$ SC is layered, that is if the system consists of two-dimensional layers, see Fig. 3.4, it could still be the case that there is a well-defined response to the two-dimensional geometry of the layers and the Wen-Zee term generalizes directly, but the flux current now corresponds to a current of strings and b^{ext} is a two-form gauge field. The promising fact is that almost all but one of the mentioned chiral SC candidates are strongly layered materials in the sense that the mechanism responsible for the superconductivity seem to take place in two-dimensional layers. This makes the possibility for a non-trivial geometric Meissner effect most probable.

As a side note. The evidence of chiral superconductivity for all these the mentioned materials is only indirect. The superconducting phase break time-reversal

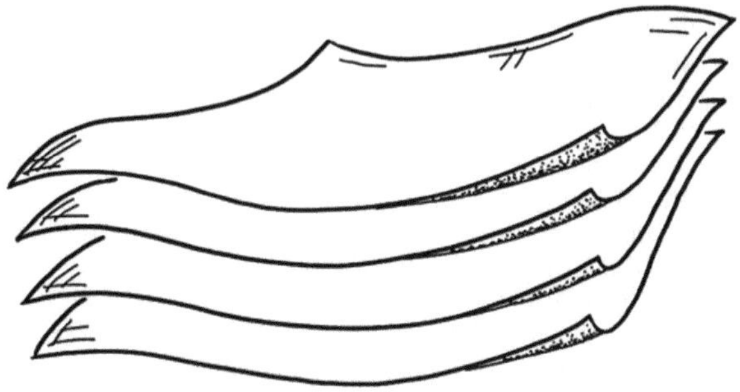

Fig. 3.4 A layered $3\mathscr{d}$ system will depend on the metric of the $2\mathscr{d}$ layers. *Figure by S. Holst*

invariance spontaneously and it is only this spontaneus breaking of time-reversal invariance that one has tried to detect.[2] Here there is possilbly a practical application for the geometric Meissner effect. The spontaneus breaking of time-reversal invariance could of course come from other mechanisms than chiral superconductivity. But the geometric Meissner effect can only occure in chiral superconductors. Flexing a material such that it $2\mathscr{d}$ curves should in principle be possible with available experimental techniques. The only problem is if one can accomplishe a big enough curvature. What is promising is, as we already said, is that the magnetic field would be measured relative to zero. The most sensitive techniques can, when averaging over long times, measur fields in the order of $\sim$aT (see e.g., Ref. [27]). This should be compared to $\sim\mu$T which one get as the maximum geometric Meissner field from a back of the envelope caluclation if one assumes a maximum bond stretching of 1% and a London length, $\lambda_L = 1\,\mu$m. (The field grows inversly proportional to λ_L^2 and $1\,\mu$m is roughly the maximum known penetration depth of any SC; for a chiral superconductor with shorter penetration depth the geometric Meissner field could possibly be even bigger.)

References

1. A. Chan, T.L. Hughes, S. Ryu, E. Fradkin, Phys. Rev. B **87**, 085132 (2013)
2. X.G. Wen, A. Zee, Phys. Rev. B **46**, 2290 (1992a)
3. K.V. Klitzing, G. Dorda, M. Pepper, Phys. Rev. Lett. **45**, 494 (1980)
4. Q. Niu, D.J. Thouless, Y.-S. Wu, Phys. Rev. B **31**, 3372 (1985)
5. R.B. Laughlin, Phys. Rev. B **23**, 5632 (1981)
6. D.J. Thouless, M. Kohmoto, M.P. Nightingale, M. den Nijs, J. Phys. Rev. Lett. **49**, 405 (1982). https://doi.org/10.1103/PhysRevLett.49.405
7. J.E. Avron, R. Seiler, B. Simon, Phys. Rev. Lett. **51**, 51 (1983). https://doi.org/10.1103/PhysRevLett.51.51
8. M. Kohmoto, Anna. Phys. **160**, 343 (1985). https://doi.org/10.1016/0003-4916(85)90148-4
9. S. Ryu, J.E. Moore, A.W.W. Ludwig, Phys. Rev. B **85**, 045104 (2012). https://doi.org/10.1103/PhysRevB.85.045104

[2] For Sr_2RuO_4 there is also experiments to detect the presence of half-quantum vortices (see e.g., [25]) but these are not conclusive, see Ref. [26].

10. C. Hoyos, D.T. Son, Phys. Rev. Lett. **108**, 066805 (2012). https://doi.org/10.1103/PhysRevLett.108.066805
11. X.G. Wen, A. Zee, Phys. Rev. Lett. **69**, 953 (1992b). https://doi.org/10.1103/PhysRevLett.69.953
12. T. Kvorning, T.H. Hansson, A. Quelle, C.M. Smith, Phys. Rev. Lett. **120**, 217002 (2018). https://doi.org/10.1103/PhysRevLett.120.217002
13. S. Moroz, A. Prem, V. Gurarie, L. Radzihovsky, Phys. Rev. B **95**, 014508 (2017). https://doi.org/10.1103/PhysRevB.95.014508
14. A. Kitaev, Ann. Phys. **321**, 2 (2006). J. Spec. Issue. https://doi.org/10.1016/j.aop.2005.10.005
15. H. Tou, Y. Kitaoka, K. Ishida, K. Asayama, N. Kimura, Y. Onuki, E. Yamamoto, Y. Haga, K. Maezawa, Phys. Rev. Lett. **80**, 3129 (1998)
16. M. Nishiyama, Y. Inada, G.-Q. Zheng, Phys. Rev. Lett. **98**, 047002 (2007)
17. Y. Maeno, S. Kittaka, T. Nomura, S. Yonezawa, K. Ishida, J. Phys. Soc. Jpn **81**, 011009 (2012)
18. M.H. Fischer, T. Neupert, C. Platt, A.P. Schnyder, W. Hanke, J. Goryo, R. Thomale, M. Sigrist, Phys. Rev. B **89**, 020509 (2014)
19. Y. Nishikubo, K. Kudo, M. Nohara, J. Phys. Soc. Jpn **80**, 055002 (2011)
20. M.L. Kiesel, C. Platt, W. Hanke, D.A. Abanin, R. Thomale, Phys. Rev. B **86**, 020507 (2012)
21. A.M. Black-Schaffer, C. Honerkamp, J. Phys. Condens. Matter **26**, 423201 (2014)
22. R. Nandkishore, L.S. Levitov, A.V. Chubukov, Nat. Phys. **8**, 158 (2012)
23. M.L. Kiesel, C. Platt, W. Hanke, R. Thomale, Phys. Rev. Lett. **111**, 097001 (2013)
24. L. Keldysh, Soviet. J. Exp. Theor. Phys. Lett. **29**, 658 (1979)
25. J. Jang, D.G. Ferguson, V. Vakaryuk, R. Budakian, S.B. Chung, P.M. Goldbart, Y. Maeno **331**, 186 (2011)
26. X. Cai, Y.A. Ying, N.E. Staley, Y. Xin, D. Fobes, T.J. Liu, Z.Q. Mao, Y. Liu, Phys. Rev. B **87**, 081104 (2013)
27. D. Robbes, Sens. Actuators Phys. **129**, 86 (2006), eMSA (2004)

Chapter 4
Topological Geometric Response and Topological Order of Super Conductors

Different aspects of topological properties of superconductors is the topic of the three of the research papers published during my time as a Ph.D student. This prominent appearance of superconductivity should not come as a surprise. Superconductivity is such a common low-temperature phase of matter that it was discovered, by Heike Kamerlingh Onnes in 1911, almost immediatly after he invented the helium-temperature cryostate. The type of super conductor discovered then was the fully gapped s-wave type, which is one of the simplest examples of topological order, see References [1, 2]. In such $3\mathscr{d}$ supoer conductors, the string-like vortex excitations and fermionic excitations (screened electrons) interact topologically; a Berry phase of -1 is generated every time a fermion encircles a vortex.

We will not study the $3\mathscr{d}$ superconductors, but instead the $2\mathscr{d}$ version, where the vortices are point particles. For the s-wave case this might seem like an unnecessary theoretical exercise, but the relevance will become clear when we discuss the chiral superconductors.

In the s-wave case there are four anyons: the trivial, the vortex, the Bogoliubov de Gennes (BdG) particle and the fusion of the two. While in the chiral p-wave SC, the fusion of a vortex and a BdG particle does not produce an additional anyon. This can, as shown later in this chapter, be seen directly from an adiabatic study of the BdG wave functions.

We will begin by studying a bosonic theory which should be thought of as a theory for the Cooper pairs in a SC. In doing this we will also use the opportunity to take a closer look at the geometric Meissner effect. We will study the geometric response of the bosonic sector of a chiral p-wave SC and derive the Wen-Zee term.

Then, we move on to investigate the fermionic sector in a simple model of spinless fermions in the mean-field BdG picture. Here we will see how one can determine the topological order by an adiabatic argument. We end this chapter by introducing the Majorino conundrum from paper 4 and 5 and discuss its solution.

T. Klein Kvorning, *Topological Quantum Matter*, Springer Theses,
https://doi.org/10.1007/978-3-319-96764-6_4

4.1 A Superconducting Bosonic Model

In this section we will consider a charged bosonic model which should be thought of as describing the Cooper pair field. We will take a somewhat more microscopic view point than in the last parts of the previous chapter, but we will still look at a phenomenological fluid model with the same symmetries as the underlying geometry. We would expect this to be a correct effective theory at long distances. In any case, what we are interested in are the topological properties that do not depend on any microscopic details—we want to understand how the topological properties can emerge.

This first section is mostly a warm up for the later sections, where these results will be used. It will give some of the technical background to the emergence of the topological field theories and response theories that will be used in later sections.

The starting point is a charge-two scalar bosonic field Φ, coupled to electromagnetism. (Below we will also study the situations where the Cooper pairs also have some internal motion and the bosonic field is vector valued.) As before, we assume a universal time direction and use h to denote the, possibly time-dependent, spatial $2d$ metric. In this section we will also make use of space-time metrics. The usual Lorentz space-time metric will be denoted $g_c = -c^2 dt \otimes dt + h$ (c is the speed of light). The theory is described by the action

$$S = \int dt \wedge \left(\star_h \Phi^\dagger (\partial_t + i2A_t)\Phi - \frac{1}{2m}\Phi^\dagger (d - i2A) \star_h (d + i2A)\Phi \right) + V(|\Phi|) + \frac{1}{2\mu_0}\int dA \wedge \star_c dA, \tag{4.1}$$

where $V(|\Phi|)$ is a potential which will remain unspecified at the moment and $\star_h$ and $\star_c$ are the Hodge duals defined by the metrics h and g_c respectively. We assume that $V(|\phi|)$ has a form which ensures mean-field solutions $\Phi = \sqrt{\bar{\rho}}\xi$, where $\bar{\rho}$ is a real constant and ξ is a singular phase which encodes vortices. In other words, ξ winds clockwise around each vortex and anti-clockwise around each anti-vortex. The integral

$$\int_\Gamma \xi^* d\xi \tag{4.2}$$

is the change of the phase of ξ along the curve Γ, and when Γ is a closed curve

$$\frac{1}{2\pi i}\int_\Gamma \xi^* d\xi = N_{vor.}, \tag{4.3}$$

where $N_{vor.}$ denotes the number of vortex world-lines, minus the number of anti-vortex world-lines encircled by Γ. Using Stokes theorem we can thus conclude that

$$\frac{1}{2\pi i}\int_S d\xi^* d\xi = N_{vor.}, \tag{4.4}$$

where S is a surface with Γ as its boundary. So, the above integral gives the number of vortices on a spatial surface, or the number of vortices passing through a non-spatial surface, i.e., the vortex current two-form is

$$\mathcal{J}_{vor.} = \frac{1}{2\pi i} d\xi^* d\xi. \tag{4.5}$$

We expand around the mean-field solution and write

$$\Phi = \left(\sqrt{\bar{\rho}} + \frac{\delta\rho}{2\sqrt{\bar{\rho}}}\right)\xi e^{i\phi}. \tag{4.6}$$

Keeping terms to second order in $\delta\rho$ and integrating out $\delta\rho$, results in

$$S = \int \frac{1}{2\mu_0} \star_c dA \wedge dA + \frac{vk}{4c}\left(\frac{1}{v^2}\star_c \left(\partial_t\phi + 2A_t + i\xi^*\partial_t\xi\right)^2 \right.$$
$$\left. - cdt \wedge (d\phi + 2A + i\xi^* d\xi) \wedge \star_h (d\phi + 2A + i\xi^* d\xi)\right), \tag{4.7}$$

where k and v are constants defined by m and the potential $V(|\phi|)$. To linearize this action we now introduce a Hubbard-Stratonovich two-form field $\mathcal{J}$ and we get

$$S = \int \mathcal{J} \wedge \left(d\phi + 2A + i\xi^* d\xi\right) + \frac{1}{k}\mathcal{J} \wedge \star_v \mathcal{J} + \frac{1}{2\mu_0}\star_c dA \wedge dA, \tag{4.8}$$

where $\star_v$ is the Hodge dual defined by the metric $g_v = -v^2 dt \otimes dt + h$. Integrating out ϕ reveals that $\mathcal{J}$ is a conserved current, $d\mathcal{J} = 0$, so we can write $\mathcal{J} = db/2\pi$ and we end up with

$$S = \int \frac{1}{k}\star_v db \wedge db + \frac{1}{2\mu_0}\star_c dA \wedge dA + \frac{1}{\pi} A \wedge db + b \wedge \mathcal{J}_{vor.}. \tag{4.9}$$

The source-free equations of motion are

$$d \star_v d \star_c dA - \frac{\sqrt{\mu_0 k}}{4\pi^2} dA = 0 \tag{4.10}$$

$$d \star_c d \star_v db - \frac{\sqrt{\mu_0 k}}{4\pi^2} db = 0, \tag{4.11}$$

so there is a gapped spectrum. We thus have a well-defined topological limit given by the terms in the action (4.8) that do not depend on any metric:

$$S = \int \frac{1}{\pi} A \wedge db + b \wedge \mathcal{J}_{vor.}. \tag{4.12}$$

Let us now see how this changes for a chiral SC with the chiral p-wave SC as an example. Now, we no longer have a scalar order parameter, but

$$\Phi = \overbrace{\frac{\sqrt{\rho_R(x)}}{\sqrt{2}} \xi_R e^{i\phi_R(x)} (e_1 + ie_2)}^{\Phi_{e_1+ie_2}} + \overbrace{\frac{\sqrt{\rho_L(x)}}{\sqrt{2}} \xi_L e^{i\phi_L(x)} (e_1 - ie_2)}^{\Phi_{e_1-ie_2}}, \tag{4.13}$$

where e_1 and e_2 are spatial unit co-vectors, i.e., $h = e_1 \otimes e_1 + e_2 \otimes e_2$. There are two independent kinds of vortices since ξ_L and ξ_R both can wind. We will assume that there is a mean-field solution with

$$\rho_R = \bar{\rho} \qquad ; \qquad \rho_L = 0. \tag{4.14}$$

We will do as before and expand around this mean-field solution. The fluctuation in ρ_L will not contribute to lowest order, so it will be ignored, and we write

$$\Phi = \frac{1}{\sqrt{2}} \left(\sqrt{\bar{\rho}} + \frac{\delta\rho}{2\sqrt{\bar{\rho}}} \right) e^{i\phi} \xi \, (e_1 + ie_2). \tag{4.15}$$

Now, since Φ is not scalar there are several choices for the Laplace operator that all preserve the symmetry of h, but they all give the same result to lowest order in the curvature and derivatives. The difference from the scalar case is that we pick up the spin connection when the derivatives act on e_1 and e_2, and we end up with

$$S = \int \mathcal{J} \wedge \left(d\phi + 2A + i\xi^* d\xi + \omega \right) + \frac{1}{k} \mathcal{J} \wedge \star_v \mathcal{J} \tag{4.16}$$

instead of (4.8), and thus

$$S = \int \frac{1}{k} \star_v db \wedge db + \frac{1}{2\mu_0} \star_c dA \wedge dA + \frac{1}{\pi} (A + A^{ext.} + \omega) \wedge db + b \wedge \mathcal{J}_{vor.}. \tag{4.17}$$

To get the partition functional which generates the electric and flux currents, we need to introduce background fields $A^{ext.}$ and b^{ext}, which couple to db and dA respectively. Adding those fields and integrating out A and b we get

$$-i \log \mathcal{Z}[A^{ext}, b^{ext}, \omega] = \frac{1}{\mu_0 k} \int d(A^{ext.} + \omega) \wedge \star_c d \star_v db^{ext} + \frac{1}{\pi} \int (A^{ext.} + \omega) \wedge db^{ext}. \tag{4.18}$$

4.2 Adding Fermions to Get Topological Order

Let us the begin with the s-wave case. The bosonic Cooper pair field technically originates as a Hubbard-Stratonovich field, used to linearize a short-range attractive interaction between electrons. Considering the simplest short-range interaction, namely the contact interaction $\propto \int \psi_\uparrow^\dagger \psi_\downarrow^\dagger \psi_\downarrow \psi_\uparrow$, we would get a fermionic theory

$$H = \int \frac{1}{2m} \sum_\sigma \psi_\sigma^\dagger (d \star_h d - \star_h \mu) \psi_\sigma + \star_h \Phi^\dagger \psi_\uparrow \psi_\downarrow, \tag{4.19}$$

(and possibly also spin mixing terms) to lowest order. Here $\uparrow$ and $\downarrow$ denote different species of fermions, e.g., different spin species. Inserting the mean-field solution for Φ there will be a gap to all fermion excitations. So, except that they do carry a conserved quantum number (i.e., the fermion number), they do not alter the low-energy theory. If a fermion is created and localized at some point, it cannot decay to vacuum without another fermion coming close. We only have to keep track of this conserved charge. So, we can model their behavior by a source term,

$$\int A \wedge \mathcal{J}_q, \tag{4.20}$$

added to the action (4.12). The resulting action describes topological order with a mutual statistics between the fermion and the vortices: braiding a fermion and a vortex adds a Berry phase -1. If the fermion would not have been there, the action would have taken the same form, but it would not have described topological order. In that case the gauge fields would not have been normalized according to the conventions in this thesis, since there would be no excitations corresponding to the Wilson loop

$$W = e^{i \int_\Gamma A}; \tag{4.21}$$

there would only be Wilson loops of the type

$$W = e^{i2 \int_\Gamma A}, \tag{4.22}$$

meaning there is no topological order.

If one studies the response functional, one would see that the fermion excitations carry no electric charge; they are perfectly screened by the Cooper pairs. But the Cooper pairs have charge two and the original electrons had charge one. So, the Cooper pairs cannot fully screen the interaction with the vortices. What is left is the minus sign obtained when encircling a vortex.

At this point one might ask: what about the weak and strong pairing phase, and type I and type II SC's? Are they not different phases? For s-wave superconductors there is no phase transition between these types only crossovers. The difference between type I and type II SC is that in the first case vortices attract, and in the second case they repel. But in both cases, the vortices only have short-range interaction, so it costs a finite amount of energy to create a vortex, anti-vortex pair and separating them to infinity. So, they are topological excitations present in the low-energy spectrum.

The difference between the weak and strong pairing phase is analagous, though the distinguishing feature is whether there is an attractive or a repulsive interaction between the fermions rather than the vortices.

4.2.1 The Chiral p-Wave SC

We will now see how the Majorino mode emerges. We will use the simplest possible model and assume only one species of fermions. We also assume that the boson comes from a Hubbard-Stratonovich decomposition of some short-range two-body interaction. We take the simplest potential that is a function of only the geodesic distance, i.e., $V(X,Y) = \lambda \star_h d \star_h d\delta^2(X-Y)$. (Note that the interaction $\delta^2(X-Y)$ is identically zero for one-species fermions.)

Using integration by parts this can be written as

$$\hat{V} = \lambda \int \psi^\dagger d\psi^\dagger \wedge \star_h \psi d\psi. \tag{4.23}$$

Plugging in the ansatz (4.14) from the previous section results in the following fermionic Hamiltonian:

$$H = \int \star_h \Psi^\dagger \begin{pmatrix} h_0 & \Phi_{e_1+ie_2}\partial_- \\ \Phi^*_{e_1+ie_2}\partial_+ & -h_0^* \end{pmatrix} \Psi, \tag{4.24}$$

where $\Psi = \left(\psi, \psi^\dagger\right)^T$, $\partial_\pm = e^{1\mu}\partial_\mu \pm ie^{1\mu}\partial_\mu$ and the non-interacting part of the Hamiltonian h_0 is yet to be specified.

We are only interested in long-wavelength effects, so we only keep the lowest order in h_0—i.e., a constant $h_0 \equiv \mu$. The above Hamiltonian seems to be symmetric with respect to changing the sign of μ. However, in case of a boundary (or singularities i.e., vortices) there is not a unique boundary condition which makes the Hamiltonian self-adjoint. Specifying this boundary condition breaks the *apparent* symmetry of changing the sign of μ. This might seem like a technical point, but since there appears to be some confusion about this in the literature, let us discuss it in some detail.

We use a first-quantized language where the Hamiltonian $\mathcal{H}$ is defined by how it acts on a general single particle state,

$$i\hbar\frac{\partial}{\partial t}\begin{pmatrix} u \\ v \end{pmatrix} = \mathcal{H}\begin{pmatrix} u \\ v; \end{pmatrix} \qquad |u,v\rangle = \int \star_h \left(\psi^\dagger, \psi\right)(u,v)^T |GS\rangle. \tag{4.25}$$

We get

$$\mathcal{H} = \begin{pmatrix} -\mu & \frac{1}{2\sqrt{h}}\{\sqrt{h}\Phi_{e_1+ie_2}, \partial_-\} \\ -\frac{1}{2\sqrt{h}}\{\sqrt{h}\Phi^*_{e_1+ie_2}, \partial_+\} & \mu \end{pmatrix}, \tag{4.26}$$

when inserting the second-quantized Hamiltonian (4.24) into the definition (4.25) (here, $\{\cdot, \cdot\}$ denotes the anti-commutator). If we have edges or singularities, the geometry is not on its own enough to pick a boundary condition which makes the Hamiltonian self-adjoint (i.e., the Hamiltonian is not essentially self-adjoint).

If we do not consider a system with some flow, i.e., that fermions exit at one place and enter at another, we need to consider boundary conditions where the probability current vanishes locally across the boundary. The most general local boundary condition for which this Hamiltonian is self-adjoint reads

$$\hat{n}^\mu e^a_\mu \sigma_a \begin{pmatrix} \Phi^*_{e_1+ie_2} u \\ \Phi_{e_1+ie_2} v \end{pmatrix}\Bigg|_{\partial S} = s|\Phi_{e_1+ie_2}| \begin{pmatrix} u \\ v \end{pmatrix}\Bigg|_{\partial S}, \tag{4.27}$$

where s is an arbitrary real number, $\hat{n}^\mu$ is the outward directed normal, and ∂S is the edge of the surface S on which the system is defined.

The first-quantized Hamiltonian satisfies

$$\sigma_x \mathcal{H}^* \sigma_x = \mathcal{H}, \tag{4.28}$$

which reflects that in first-quantized language, we formally have doubled the degrees of freedom. This property of the first-quantized Hamiltonian, is required for $\psi^\dagger$ to be the adjoint of ψ, and is thus not a symmetry, but a consequence of the second-quantized structure. The only boundary conditions consistent with (4.28) are the ones with unit s. Thus, we only have two choices:

$$s = \pm 1. \tag{4.29}$$

One way to make sense of these two choices is to consider what would happen if we added higher derivative terms. If we take the natural single-particle Hamiltonian, $h_0 = - \star_h d \star_h d/(2M) - \mu$ in (4.24), and take the $M \to \infty$ limit, we will re-obtain the Hamiltonian in (4.26), together with the boundary condition $s = 1$. So, higher order terms pick a boundary condition and thus remove the apparent symmetry of changing the sign of μ.

4.2.1.1 The Majorino Modes

In this section we will demonstrate how the Majorino modes appear. Modes localized at vortices or boundaries, with an energy smaller than the bulk gap, are present for generic superconductors, see Ref. [3], but for odd chiral superconductors there is one mode that is special: a zero-dimensional, localized Majorana—*the Majorino.*

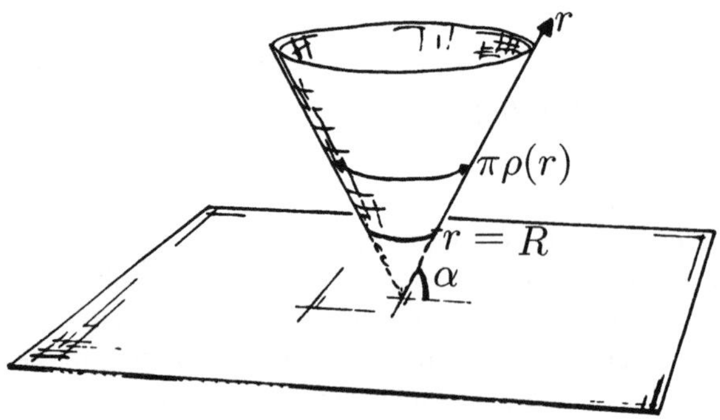

Fig. 4.1 The conical geometry which interpolates between the annulus and the cylinder. For $\alpha = 0$ we have an annulus, and for $\alpha = \pi/2$ we have a cylinder. R is the radius of the cone where it is cut, so for $\alpha = 0$, r is just the usual radial coordinate. The radius of the cone at an arbitrary radius is denoted $\rho(r) = (1 - \cos(\alpha))R + r\cos(\alpha)$

We will study both the vortex and edge Majorinos and the chiral edge Majorana in one go. For simplicity, we assume a homogeneous system with rotation invariance around a fixed axis, i.e., a cone with the top cut out, see Fig. 4.1. We will use polar coordinates (r, θ), where r is defined in Fig. 4.1 and θ is the perpendicular direction, with $\theta \equiv \theta + 2\pi$. We have rotation invariance, so without loss of generality we can assume

$$\Phi_{\hat{r}+i\hat{\theta}} = e^{-im\theta}\Delta(r), \tag{4.30}$$

where m is an integer. We will also, for simplicity, assume that the superconducting coherence length is much smaller than all other relevant length scales, i.e., we put $\Delta(r) \equiv \Delta = const$. This does not make any difference when it comes to any results, but it streamlines the calculations.

A direct calculation yields

$$\frac{1}{2\sqrt{h}}\{\sqrt{h}\Phi_{\hat{r}+i\hat{\theta}}, \partial_-\} = e^{-im\theta}\Delta\left(\partial_r + \frac{i}{\rho(r)}\partial_\theta + \frac{m+\cos\alpha}{2\rho(r)}\right),$$

where $\rho(r)$ and α is defined in Fig. 4.1. Substituting the ansatz

$$\begin{pmatrix} u \\ v \end{pmatrix} = e^{il\theta}\begin{pmatrix} e^{-im\theta/2}u_l(r) \\ e^{im\theta/2}v_l(r) \end{pmatrix} \quad ; \quad l \in \mathbb{Z}, \tag{4.31}$$

into the Hamiltonian (4.26), and eliminating v_l, we obtain

$$\left[\mu^2 - E^2 - \Delta^2\left(\partial_r - \frac{l-\frac{1}{2}}{r}\right)\left(\partial_r + \frac{l+\frac{1}{2}}{r}\right)\right]u_l(r) = 0. \tag{4.32}$$

The boundary boundary condition (4.27) is $v_l(R) = s u_l(R)$; or, expressed solely in terms of u_l (by using the equations of motion):

$$\lim_{r \to R} \left[s(\mu - E) + \Delta \left(\partial_r + \frac{l + \frac{1}{2}\cos\alpha}{r} \right) \right] u_l(r) = 0. \tag{4.33}$$

Let us now have a closer look at (4.32). For $|E| > \mu$ there are plane-wave solutions, so μ is the bulk energy gap. From (4.32), we also see that for $|E| < \mu$ there are two solutions: one which decays with increasing r, and one which increases. Which solution is allowed by the boundary condition depends on the sign of μs; the decaying solution is only allowed for positive μs and the increasing is only allowed for negative μs. Since the solutions have to be normalizable, this reflects that when μ changes sign, the system undergoes a phase transition from a phase that supports edge modes to a phase that does not, or *vice versa*.

In the limit $R \to \infty$, and for $\mu s > 0$, the expression for the full spectrum takes the simple form

$$u_l(r) \sim \exp\left[-\left(\frac{s\mu}{\Delta} + \frac{\cos\alpha}{2R_i} \right) r \right] \quad ; \quad E_l = \frac{\Delta}{R_i} l s, \tag{4.34}$$

which is a chiral edge Majorana.

Let us now study the exact zero-energy mode in more detail. From (4.28) it follows that the spectrum is symmetric around $E = 0$. This means that if we have an odd number of discrete (and thus localized) modes, we must have an odd number of zero-modes. Changing the parameters, modes can come down from the continuum and become discrete, and *vice versa*. But, because of (4.28), they must always come in pairs. If we have one zero-mode, we will continue to have at least one as long as we do not close the bulk gap. If there, by chance, are two zero-modes, a small perturbation would in general gap them out, and for generic parameters we therefore expect to have exactly one zero-mode. We have considered a simplified model, but as long as we can interpolate between this model and a realistic one, without closing the bulk energy gap, the simplified model will predict the same number of zero-modes.

For $m \in 2\mathbb{Z}$, there is a mode with $l = m/2$, $E = 0$, and by a gauge transformation we can, for $m \in 2\mathbb{Z}$, put $m = 0$. Then, as long as μs is positive, we have the single exact solution for any value of the cone opening angle α,

$$u_0(r) = v_0(r) \propto \frac{1}{\sqrt{\rho(r)}} e^{-s\mu r/\Delta}. \tag{4.35}$$

Let us now take $R \to 0$. (The edge Majorinos will be discussed later.) From the previous section we remember that the covariant derivative on $\Phi_{\hat{r}+i\hat{\theta}}$ is

$$(d + i2A + i\omega)\, \Phi_{\hat{r}+i\hat{\theta}}, \tag{4.36}$$

and in polar coordinates on a plane we have $\omega = d\theta$. The precise form of the kinetic energy functional does not matter; in any case, a larger covariant derivative will mean a larger kinetic energy. If the solution with $m = 0$ should not cost infinite energy, we must have the asymptotic behavior $2A \to -d\theta$ when $r \to \infty$. That is, there has to be a superconducting vortex. The $m = 0$ solution thus corresponds to the fundamental vortex solution and close to the vortex we will have a zero-mode localized with the localization length $\lambda_Z = |\Phi_{\hat{r}+i\hat{\theta}}|/\mu$. Note that this length scale is independent of the SC coherence length, which we have put to zero. We can now finally conclude that we will have a fermionic mode

$$\gamma = c^\dagger + c, \tag{4.37}$$

where $c^\dagger$ is the localized mode $c^\dagger = \int \star_h \psi^\dagger u_0$ (the definition of u_0 is given in Eq. (4.35)), that commutes with the Hamiltonian.

Let us next consider a situation with several vortices and use Cartesian coordinates (x_1, x_2). Then the spin connection vanishes, and the solution for the bosonic field becomes

$$\Phi_{\hat{x}_1+i\hat{x}_2}(X) = e^{i\sum_i \arg(X-X_i)}\Delta(X). \tag{4.38}$$

When solving for the fermion modes, we can treat the vortices as isolated if the minimal separation distance between them is much larger than λ_Z. So, the zero-mode operator at vortex i is

$$\gamma_i = e^{i\chi_i}c_i^\dagger + e^{-i\chi_i}c_i, \tag{4.39}$$

where $c_i^\dagger$ is the operator which creates the local mode $c_i^\dagger = \int \star_h \psi^\dagger u_0(|X - X_i|)$ and $e^{i\chi_i}$ is a square root of the phase $e^{i\Omega_i} \equiv e^{i\sum_{j\neq i}\arg(X_i-X_j)}$, i.e.,

$$e^{i2\chi_i} = e^{i\Omega_i}. \tag{4.40}$$

We can pair up these $2N$ Majorino modes to form N non-local fermion modes,

$$f_i = \frac{1}{2}(\gamma_i + i\gamma_{2N-i}). \tag{4.41}$$

One ground state, $|\mathbf{0}\rangle$, is annihilated by all f_i, i.e.,

$$|\mathbf{0}\rangle \propto \prod_i f_i|GS\rangle, \tag{4.42}$$

where $|GS\rangle$ is any ground state of the Hamiltonian which is not annihilated by the f_i's. The other ground states $\{|\mathbf{k}\rangle\}$ can then be formed by acting with the different $f_i^\dagger$:

$$|\mathbf{k}\rangle = \left(f_0^\dagger\right)^{k_0}\cdots\left(f_N^\dagger\right)^{k_N}|\mathbf{0}\rangle \tag{4.43}$$

with $k_i = 0, 1$. Note that one cannot distinguish the different states $|\mathbf{k}\rangle$ locally— they are all equal superpositions of having the state created by $c_i^\dagger$ occupied or not. If one wants to tell them apart, one would have to make a joint measurement on separated vortices.

4.3 Non-Abelian Statistics in the Adiabatic Limit

The history of non-Abelian statistics of the chiral p-wave SC goes back to Ref. [4]. There, a rewriting of the first proposed non-Abelian fractional quantum Hall state as a (Cooper) paired state was suggested. An important clarification was made in Ref. [5] and, the year after, in Ref. [6] an argument, later strengthened in Ref. [7], was given for why the vortices in real chiral p-wave superconductors are non-Abelian anyons.

With this said, we now proceed by an adiabatic analysis to determine the topological interaction of the vortices in the previous section. But let us begin with a short review of the notion of geometric phases in the adiabatic limit. Consider any quantum system which starts out in a state that is part of a set of degenerate states, with a splitting ϵ and a gap Δ to all other states. We then assume that time evolution is fast compared to $\hbar/\epsilon$ but slow compared to $\hbar/\Delta$. Then, the time evolution operator (with the dynamic phase $e^{i\int^t E_{GS}(t')dt'}$ removed) is given by the Kato operator

$$\mathcal{K}_\Gamma = pe^{-i\int_\Gamma a}, \tag{4.44}$$

see Ref. [8]. Here, Γ is the path followed in parameter space, p denotes path ordering, $a = [P, dP]$ and P is the projector to the degenerate subspace.

In a specific basis $\{|\alpha(X)\rangle\}$, that depends smoothly on the parameters X, the representation of a is

$$a \doteq [\mathbb{A}]_{\alpha\beta}(X) = \Im \langle \alpha(X)| d |\beta(X)\rangle , \tag{4.45}$$

where $\Im$ denotes the imaginary part and d is the exterior derivative in parameter space. If the parameter space has a non-trivial fundamental group, we can (as in Eq. (2.5)) absorb some of the connection by having a multivalued basis $\{|\alpha(X)\rangle\}$.

To make this less abstract, we can, as an exercise, consider the example of two identical fermions localized at some position X_1 and X_2. We let Γ be a path where we have interchanged these fermions and $\mathcal{K}_\Gamma$ thus will pick up an extra minus sign from the interchange. If we have smooth basis vectors $|\alpha(X_1, X_2)\rangle$, we get this sign directly from (4.45). On the other hand, if we do as we did to get to (2.5)—i.e., we let $|\alpha(X_1, X_2)\rangle$ be an odd function on the double cover, $\mathcal{M}^2$, of $\mathcal{M}^2/S_2$ (i.e., it picks up a minus sign when X_1 and X_2 interchange)— the expression (4.45) will vanish.

The conclusion we should draw is: to get the full representation of the Kato operator we, should add the phase we get from analytically continuing the basis to the basis dependent expression (4.45). These two basis dependent parts of the Kato operator—the explicit phase from analytic continuation and the phase from (4.45)—are respectively referred to as the monodromy of the basis and the Berry phase.

With the above general discussion, we are ready to calculate the monodromy of the basis (4.43). We assume that we have $2N$ vortices that are pinned to potentials, and the position of these potentials are the parameters we vary. As we learnt in Chap. 2, the fundamental group is generated by transpositions, T_i, of vortex i and $i+1$. Also the representation of the Kato operator corresponding to the interchange T_i will be denoted $\tau(T_i)$ and the monodromy part we will denote by $\tilde{\tau}(T_i)$.

The monodromy has two contributions when interchanging vortices. For instance, when we interchange vortex 0 and 1 we first see that the operator c_0 in Eq. (4.39) does not return to itself but goes to c_1, and, secondly we have a contribution from the phase $e^{i\chi_{0/1}}$ (see (4.40)). The path where the vortices are interchanged can be parametrized by $r = |X_0 - X_1|$ and $\theta = \arg(X_0 - X_1) - \arg(X_0^{start} - X_1^{start})$. For an exchange, r goes from r^{start} back to r^{start} and θ goes from 0 to π. To choose the square root $e^{i\chi_{0/1}}$, let us lift $e^{i\Omega_{0/1}}$ to $\Omega_{0/1}$. We can choose $0 \leq \Omega_{0/1}^{start} < 2\pi$ to get

$$\Omega_0 = \theta + \lambda_0(r,\theta) \qquad ; \qquad \Omega_1 = \theta + \lambda_1(r,\theta) + \pi, \tag{4.46}$$

where λ_i is the phase dependence from the other vortices. Since no other vortices get encircled, we get no winding in λ_i, and it returns to itself after the interchange, i.e., $\lambda_{0/1} \to \lambda_{1/0}$. In the definition of the basis $\{|\mathbf{k}\rangle\}$ we have a freedom in choosing the sign of the Majorino operators, or, put differently, when taking the square root of $e^{i\Omega_i}$ to get $e^{i\chi_i}$ we have two choices for each γ_i. In terms of $\chi_{0/1}$, the choice amounts to either choosing

$$\chi_0^{start} = \frac{\lambda_0}{2} \qquad ; \qquad \chi_1^{start} = \frac{\lambda_1 + \pi}{2} \tag{4.47}$$

or

$$\chi_0^{start} = \frac{\lambda_0}{2} \qquad ; \qquad \chi_1^{start} = \frac{\lambda_1 - \pi}{2}. \tag{4.48}$$

When θ goes from 0 to π, the basis choice (4.47) implies

$$\chi_0 \to \chi_1 \qquad ; \qquad \chi_1 \to \chi_0 + \pi, \tag{4.49}$$

which in combination with $c_{0/1} \to c_{1/0}$ gives the result

$$\gamma_0 \to \gamma_1 \qquad ; \qquad \gamma_1 \to -\gamma_0. \tag{4.50}$$

Which fermion operator that gets the minus sign under a transposition has no physical meaning, and is just an attribute of the basis choice we just made.

Using the anti-commutation relations $\{\gamma_i, \gamma_j\} = 2\delta_{ij}$, we can conclude that the operator

$$U = \frac{1}{\sqrt{2}}(1 + \gamma_1\gamma_0) \tag{4.51}$$

is unitary, has the property

$$U\gamma_{(1\pm1)/2}U^\dagger = \mp\gamma_{(1\mp1)/2} \tag{4.52}$$

and commutes with $\gamma_{0/1}$— as well as any operator that commutes or anti-commutes with $\gamma_{0/1}$. We thus see that U generates the change (4.50). So, when we analytically continue

$$\mathcal{O}_{\mathbf{k}} \equiv \left(f_0^\dagger\right)^{k_0} \cdots \left(f_N^\dagger\right)^{k_N} \tag{4.53}$$

along a path corresponding to T_0, we get $U\mathcal{O}_{\mathbf{k}}U^\dagger$. From this we conclude that when we analytically continue $|\mathbf{k}\rangle$ along the same path, we get

$$U|\mathbf{k}\rangle = U\mathcal{O}_{\mathbf{k}}U^\dagger U|0\rangle \tag{4.54}$$

up to a possible phase generated when U acts on $|0\rangle$ (recall the definition of $|\mathbf{k}\rangle$ from Eq. (4.43)). We thus have

$$\tilde{\tau}(T_0) = U = \frac{e^{i\phi}}{\sqrt{2}}(1 + \gamma_1\gamma_0), \tag{4.55}$$

where ϕ is some yet unspecified phase. For a general transposition, T_i, we can just replace γ_1 with γ_{i+1}, and γ_0 with γ_i.

We are now left with calculating the Berry phase. First we need to get an expression for the exterior derivative acting on γ_i.

We note that the operators γ_i depend on the positions of the vortices both through the wave function u_0 and the phase $e^{i\chi_i}$, recall the expression

$$\gamma_i = e^{i\chi_i}c_i^\dagger + e^{-i\chi_i}c_i, \tag{4.56}$$

where

$$c_i^\dagger = \int \star_h\psi^\dagger(X)u_0(|X - X_i|) \tag{4.57}$$

and u_0 is exponentially decaing away from X_i. When the exterior deriviative act on γ_i we get one term

$$\left(e^{i\chi_i}c_i^\dagger - e^{-i\chi_i}c_i\right) id\chi_i \equiv \tilde{\gamma}_i d\chi_i \tag{4.58}$$

from when the derivative acts on χ_i. We can, by a direct calculation, conclude that $\tilde{\gamma}_i$, defined in the above equation, anti-commutes with all Majorino modes.

We can in general write the term gotten when d acts on u_0 on the form,

$$\gamma_i dk_1 + \gamma_i^{\perp} dk_2. \tag{4.59}$$

Here $\{dk_i\}$ is some unspecified one-forms in parameter space, and $\gamma_i^{\perp}$ denotes an operator that anti-commutes with γ_i. Since u_0 is real, we can conclude that dk_1 is real, and since u_0 is exponentially localized at X_i we can conclude that $\gamma_i^{\perp}$ does not only anti-commute with γ_i, but with *all* Majorino modes.

Recalling the definition

$$f_i^{\dagger} = \gamma_i - i\gamma_{i+1} \tag{4.60}$$

we can from (4.58) together with (4.59) conclude that

$$df_i^{\dagger} = d(\gamma_i - i\gamma_{i+1}) = f_i^{\dagger} dk_2 + f_i dk_3 + \gamma_i^{\perp} dk_4, \tag{4.61}$$

where dk_2 is real and $\gamma_i^{\perp}$ denotes an operator which anti-commute with all the Majorino operators. From this we get the following expression for the Berry connection,

$$\begin{aligned}
&\Im\Big\langle \mathbf{0} \Big| (f_0)^{k_0'} \cdots (f_N)^{k_N'} \, d\left(f_0^{\dagger}\right)^{k_0} \cdots \left(f_N^{\dagger}\right)^{k_N} \Big| \mathbf{0} \Big\rangle = \\
&= \sum_i \delta_{k_i',0}\delta_{k_i,1}\delta_{k_0',k_0} \cdots \delta_{k_{i-1}',k_{i-1}}\delta_{k_{i+1}',k_{i+1}} \cdots \delta_{k_N',k_N} \Im\Big\langle \mathbf{0} \Big| df_i^{\dagger} \Big| \mathbf{0} \Big\rangle \\
&+ \sum_i \delta_{k_i'+k_i,2}\delta_{k_0',k_0} \cdots \delta_{k_{i-1}',k_{i-1}}\delta_{k_{i+1}',k_{i+1}} \cdots \delta_{k_N',k_N} \Im\Big\langle \mathbf{0} \Big| dk_2 \Big| \mathbf{0} \Big\rangle \\
&\qquad + \delta_{\mathbf{k}\mathbf{k}'} \Im\Big\langle \mathbf{0} \Big| d \Big| \mathbf{0} \Big\rangle,
\end{aligned} \tag{4.62}$$

where $\Im$ denote "imaginary part". The second term on the right-hand side in the above equation is zero, simply because dk_2 is real. The first term is also zero; even though $|\mathbf{0}\rangle$ does not have a well-defined fermion number, it *has* a well-defined *fermion parity*. The only term left is thus $\Im\delta_{\mathbf{k}\mathbf{k}'}\langle \mathbf{0}|d|\mathbf{0}\rangle$, which gives a contribution proportional to identity, i.e., an overall phase. We can thus, finally, conclude

$$\tau(T_i) = \frac{e^{i\phi}}{\sqrt{2}}(1 + \gamma_{i+1}\gamma_i). \tag{4.63}$$

To get the value of the phase ϕ, we could make a detailed calculation. Furtunalty, though, we can also get it from a simple argument. If we consider fusing two vortices, we get a double vortex. Without changing the topological properties, we can deform this double vortex locally in any way we like. For example, we can open a hole in our system and squeeze all the flux in there. We then have a full flux quantum passing through a hole outside our system. This is equivalent to having nothing, since we can remove the flux quantum by a gauge transformation and close the hole. We are thus

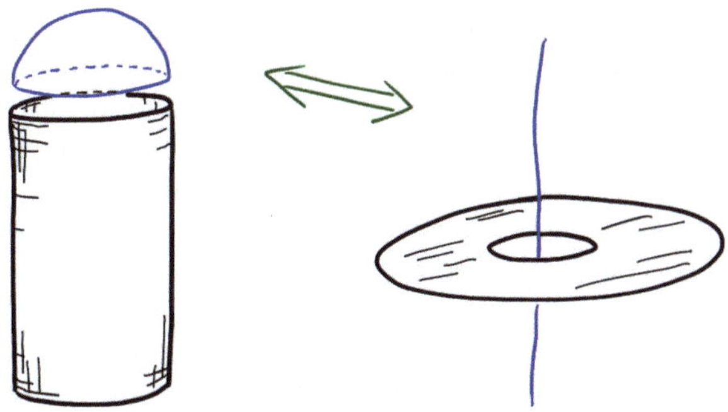

Fig. 4.2 There is a Majorino edge mode in the zero flux sector on the cylinder, since the missing spherical cap of the cylinder has the same effect as a superconducting flux-quanta. *Figure by S. Holst*

left with a defect which carries zero flux, hence either a local excitation or a fermion. There is then only one possibility which is compatible with (4.63), the fusion rules of vortices, which we denote σ, must be

$$\sigma \times \sigma = \mathbb{1} + \psi, \tag{4.64}$$

where ψ is a localized fermion (which has the same topological interactions as in the s-wave case, i.e., it gets a minus sign when encircling a vortex). This fusion rule is only compatible with $\phi = 0$, and we thus know the full representation of the infinite braid group: to get the braid tensor (2.31), we just take a four-vortex state and diagonalize $\tau(T_1)$ and $\tau(T_3)$. To get the F-tensor we also diagonalize $\tau(T_2)$ and $\tau(T_4)$ and calculate the overlaps with the basis states where $\tau(T_1)$ and $\tau(T_3)$ are diagonal.

4.4 The Majorino Conundrum

In paper 3 and 4 we discuss a conundrum concerning Majorino *edge* modes. As we shall see, this conundrum is resolved by the geometric Meissner which we introduced in Sect. 3.3 of the previous chapter. But to understand why, we must first understand what the problem is. Let us begin with the conundrum in paper 3. Recall, from the discussion when we derived the (4.35), that the only thing that matters for whether we get an edge zero-mode or not, is if m in

$$\Phi_{\hat{r}+i\hat{\theta}} \propto e^{i\theta m}\Delta(r), \tag{4.65}$$

is even or odd (with a gauge transformation, m can be taken to be zero or one).

As we concluded then, on an annalus with a hole around $r = 0$, we must have

$$2A \to -(m+1)d\theta, \tag{4.66}$$

since $\omega = d\theta$ or otherwise we would have kinetic energy all the way to infinity. So an edge Majorino is present if and only if we have a flux through the hole in the annalus. On the cylinder, on the other hand, $\omega = 0$ and we have the opposite situation, i.e., an edge Majorino present without flux. If one were to close a cylinder at one end, one would necessarily get some region of total curvature 2π. The Cooper pairs respond to this *missing* curvature in exactly the same way as it does to a flux through the hole. This is one way to see why the two situations are analogous— see Fig. 4.2.

As we discussed in the previous section, a Majorino mode is not something that can disappear without closing the bulk energy gap. But, with the cone as an intermediate step, we can continuously interpolate between a cylinder and an annalus. So, what happens to the zero-mode then? If we start with the cylinder geometry in its ground state, i.e., with no flux through the hole, we have edge Majorinos. But in the ground state of the annalus we do not. The solution to the conundrum is simply that we do not end up in the ground state!

On the cone with angle α, as in Fig. 4.1, the spin connection is

$$\omega = -\cos(\alpha)d\theta. \tag{4.67}$$

Then, because of the geometric Meissner effect, there will be flux through the hole in the ground state, and the vector potential will be

$$A = \begin{cases} \cos(\alpha)d\theta & \alpha < \pi/3 \\ (\cos(\alpha) - 1)d\theta & \alpha > \pi/3. \end{cases} \tag{4.68}$$

There is also an excited state corresponding to having an extra superconducting flux-quantum through the hole:

$$A = \begin{cases} \cos(\alpha)d\theta & \alpha > \pi/3 \\ (\cos(\alpha) - 1)d\theta & \alpha < \pi/3. \end{cases} \tag{4.69}$$

The solution is simply that the ground state at $\alpha = \pi/3 + \epsilon$ will at macroscopic scales be different from the one at $\alpha = \pi/3 - \epsilon$. So, the tunneling between them will, for

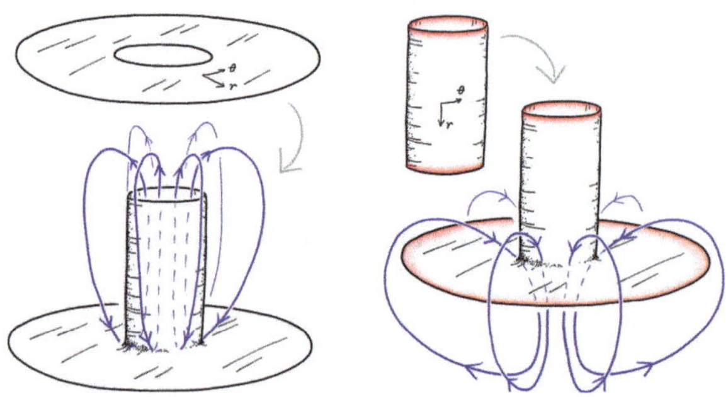

Fig. 4.3 **Left**. Lifting the inner circle of an annulus to form a cylinder. **Right**. Flattening the lower end of a cylinder to form an annulus. (Red indicates Majorino modes.)

a large enough system, vanish. If one slowly interpolates between the cylinder and the annulus, one would thus not end up in the ground state, but in the state with one superconducting flux-quantum through the hole, and this state *does* support edge Majorinos.

There are even stranger situations. For instance, what happens if we continuously lift out a cylinder from an annalus? Then, we will have a geometry with one boundary that is like a plane, and one that is like a cylinder. But the Majorinos have to come in pairs!? What happens is depicted in Fig. 4.3 and is, among other things, discussed in Ref. [9]. In Fig. 4.3 we lift the inner circle of an annulus to form a cylinder, which will create negative curvature and thus a flux density. The tunneling rate from the inner to the outer edge will be exponentially small in system size, so for a big enough system, all the flux will come from the inner edge corresponding to a flux passing through the created cylinder. On the left in Fig. 4.3 we instead flatten the lower end of a cylinder to form an annulus which will create negative curvature and thus a flux density. The tunneling rate of flux from the upper edge will be exponentially small in system size, so for a big enough system, all the flux will come from the inner edge, corresponding to a flux passing through the created annulus.

References

1. T. Hansson,V., S. Sondhi, Ann. Phys. **313**, 497 (2004). https://doi.org/10.1016/j.aop.2004.05.006
2. A. Balachandran and P. Teotontio-Sobrinho, Int. J. Modern Phys. A **08**, 723 (1993). https://doi.org/10.1142/S0217751X9300028X
3. C. Caroli, P.D. Gennes, J. Matricon, Phys. Lett. **9**, 307 (1964). https://doi.org/10.1016/0031-9163(64)90375-0
4. M. Greiter, X. Wen, F. Wilczek, Nucl. Phys. B **374**, 567 (1992). https://doi.org/10.1016/0550-3213(92)90401-V
5. N. Read, D. Green, Phys. Rev. B **61**, 10267 (2000). https://doi.org/10.1103/PhysRevB.61.10267
6. D.A. Ivanov, Phys. Rev. Lett. **86**, 268 (2001). https://doi.org/10.1103/PhysRevLett.86.268
7. A. Stern, F. von Oppen, E. Mariani, Phys. Rev. B **70**, 205338 (2004). https://doi.org/10.1103/PhysRevB.70.205338
8. T. Kato, J. Phys. Soc. Jpn **5**, 435 (1950). https://doi.org/10.1143/JPSJ.5.435
9. T. Kvorning, T.H. Hansson, A. Quelle, C.M. Smith, Phys. Rev. Lett. **120**, 217002 (2018). https://doi.org/10.1103/PhysRevLett.120.217002

the cylinder and

the annulus

support

References

Appendix A
Conventions and Notation Concerning Differential Forms and $U(1)$ Connections

This chapter is meant for definition purposes and as a short review. A thorough and more pedagoigcal explination is out of the scope of this text, but there are several good references one can consult. One of the most popular, among physicists, is Ref. [1].

A.1 Differential Forms

One-forms are fields on a manifold that specify a flow, and they can be considered as maps from curves to real numbers. (Remember that when we make reference to a curve (or other manifold or submanifold) an orientation is always implicit, so also here.) In index notation, a one-form is written as a set of real numbers, here labeled by μ, $a_\mu(X)$, where a specific basis $\{x^\mu\}_{\mu=1,\ldots,d}$ is in mind. The map, from curves to the real numbers, is

$$\Gamma \to \sum_i \int a_i \frac{dx^i(s)}{ds} ds, \tag{A.1}$$

where

$$s \to X(s) \doteq (x^1(s), x^2(s), \ldots) \tag{A.2}$$

is a parametrization of Γ. We will simply write

$$a = a_1 dx^1 + a_2 dx^2 + \cdots \tag{A.3}$$

without reference to any coordinate system, and the above equation is simply written as $\Gamma \to \int_\Gamma a$. Completely analogous, two-forms are maps from surfaces to reals and are denoted by

$$F = F_{12} dx^1 \wedge dx^2 + F_{13} dx^1 \wedge dx^3 + \cdots . \tag{A.4}$$

T. Klein Kvorning, *Topological Quantum Matter*, Springer Theses,
https://doi.org/10.1007/978-3-319-96764-6

Here, $dx^1 \wedge dx^2 = -dx^2 \wedge dx^1$ and, just as the notation implies, the wedge can be interpreted as a product of two one-forms. The map from surfaces $\{\mathcal{S}\}$ to reals is then given by

$$\mathcal{S} \to \int_{\mathcal{S}} F = \sum_{ij} \int \int_a^b F_{ij} \left(\frac{\partial x^i(s,t)}{\partial s} \frac{\partial x^j(s,t)}{\partial t} - \frac{\partial x^j(s,t)}{\partial s} \frac{\partial x^i(s,t)}{\partial t} \right) dsdt, \tag{A.5}$$

where

$$(s,t) \to X(s,t) \doteq (x^1(s,t), x^2(s,t), \dots) \tag{A.6}$$

is a parametrization of $\mathcal{S}$. Analogous definitions hold for higher-order forms.

A.2 Normalization

We will reserve the letters a and b for $U(1)$ connections. A $U(1)$ connection specifies an element in $U(1)$ for each curve. These elements are given by integrals, e.g.,

$$e^{i \int_\Gamma a} \in U(1). \tag{A.7}$$

The $U(1)$ connections could thus be thought of as one-forms up to gauge transformations

$$a \to a + i\xi^* d\xi, \tag{A.8}$$

where ξ is a function from the manifold to unit complex numbers. The usual electromagnetic vector potential is also a $U(1)$ connection and the same conventions apply. In other words we will absorb a factor of $\frac{2\pi}{\phi_0}$ in the definition of A (ϕ_0 is the flux quantum, $\phi_0 = h/e$).

Appendix B
Definitions

As a reminder: we always implicitly assume an orientation of curves, surfaces and manifolds in general. In other words, "S is a surface" is short for "S is a surface with a given orientation".

When discussing space-time we will, like in the rest of the thesis (unless otherwise specified), assume that there at least implicity is a notion of absolute rest and absolute time.

Definition 1 (*algebraic definition*) *The exterior derivative*
A lower case d denotes the exterior derivative. Acting on functions it gives a one-form,

$$df = \frac{\partial f}{\partial x^1}dx^1 + \frac{\partial f}{\partial x^2}dx^2 \dots \tag{B.1}$$

and on forms it is defined through the recursive relation

$$d(df) = 0; \qquad\qquad d(a \wedge b) = da \wedge b + (-1)^p a \wedge db.$$

Example The exterior derivative When d acts on the one-form $a = B(r)d\theta$ (where r and θ are coordinates) we get

$$da = \frac{\partial B}{\partial r}dr \wedge d\theta - B(r)d(d\theta) = \frac{\partial B}{\partial r}dr \wedge d\theta. \tag{B.2}$$

Definition 1, is usually not the best way to think of the exterior derivative. Instead, the following, equivalent definition, is better:

Definition 2 (*geometric definition*) *The exterior derivative*
The exterior derivative of a k-form a is the unique $k+1$ form da which for all $k+1$ dimensional submanifolds m uphold

$$\int_m da = \int_{\partial m} a\ . \tag{B.3}$$

T. Klein Kvorning, *Topological Quantum Matter*, Springer Theses,
https://doi.org/10.1007/978-3-319-96764-6

Definition 3 *The functional derivative*
Say that we have a local functional $S[a]$ of a k-form field a in a $\mathcal{D}$ dimensional manifold. If there exits a unique $\mathcal{D}-k$ form field $\frac{\delta S[a]}{\delta a}$ which makes the equality

$$\lim_{\epsilon\to 0}\frac{S[a+\epsilon\eta]-S[a]}{\epsilon}=\int\eta\wedge\frac{\delta S[a]}{\delta a} \tag{B.4}$$

hold for all k-form fields η, we say that the functional derivative of $S[a]$ with respect to a is that field.

Definition 4 *The Poincaré dual*
A Poincaré dual of a k-dimensional submanifold m_k is denoted by $\mathcal{P}(m_k)$ and is defined such that for all l-dimensional ($l \geq k$) submanifolds m_l and k-forms a, the following equality holds:

$$\int_{m_l}\mathcal{P}(m_k)\wedge a=\int_{m_l\cap m_k} a. \tag{B.5}$$

Lemma 5 *An explicit form of Poincaré duals*
If we consider m_k as a subset of a $\mathcal{d}$-dimensional manifold $\mathcal{M}$, it is straight forward to write an explicit form of the Poincaré duals. Say that m_k is defined implicitly by the equations $f_1(X)=f_2(X)=\cdots=f_{D-k}(X)=0$ and $f_{\mathcal{d}-k+1}(X)>0$. Then the Poincaré dual is given by

$$\mathcal{P}\left(m_k\right)=\Theta\left(f_{D-k+1}\right)\delta(f_1)\delta(f_2)\cdots df_1\wedge df_2\wedge df_3\cdots, \tag{B.6}$$

where Θ denotes the Heavyside step function. (The implicit equations do not determine an orientation for m_k but an orientation is implied by the order of the functions, $f_1, f_2, \ldots$.)

Corollary 6 *The exterior derivative of a Poincaré dual*
Using the above form of the Poincaré duals, we get

$$\begin{aligned} d\mathcal{P}\left(m_k\right)&=d\Theta\left(f_{D-k+1}\right)\delta(f_1)\delta(f_2)\cdots df_1\wedge df_2\cdots\\ &=(-1)^p\delta(f_1)\delta(f_2)\cdots df_1\wedge df_2\wedge df_3\cdots=(-1)^p\mathcal{P}\left(\partial m_k\right). \end{aligned} \tag{B.7}$$

Example 7 The linking number
In a three-dimensional manifold $\mathcal{M}$,

$$\int_{\mathcal{M}}\mathcal{P}(\Gamma)\wedge\mathcal{P}(S) \tag{B.8}$$

equals the number of times the curve Γ passes through S in the positive direction. So, the linking number between ∂S and Γ is given by the above integral.

Definition 8 (*algebraic definition*) *The Hodge isomorphism*
In a $\mathscr{d}$-dimensional oriented manifold equipped with a metric h, there is a natural map from k-forms to $\mathscr{d} - k$-forms, called the *Hodge isomorphism.* The image of a k-form a is denoted $\star_h a$ and is the unique $\mathscr{d} - k$-form, such that the equality

$$ b \wedge \star_h a = \langle b, a \rangle_h V $$

holds for all k-forms b. Above, V denotes the volume form (which is defined by the metric and the orientation), *i.e.,* written in terms of a coordinate system $\{x^\mu\}_{\mu=1,\ldots,\mathscr{d}}$,

$$ V = \sqrt{\det\left(h_{\mu\nu}\right)} dx^{\mu_1} \wedge dx^{\mu_2} \wedge \cdots \tag{B.9} $$

and $\langle\cdot,\cdot\rangle_h$ denotes the inner product between k-forms defined by h, *i.e.,*

$$ \langle b, a \rangle_h \equiv h^{\mu_1\nu_1} h^{\mu_1\nu_1} \cdots h^{\mu_k\nu_k} b_{\mu_1,\mu_2,\ldots,\mu_k} a_{\nu_1,\nu_2,\ldots,\nu_k}. \tag{B.10} $$

Definition 8 is good for calculations, but it is not the intuitive definition one should have in mind. Let us consider the Hodge isomorphism of a two-form $\mathcal{B}$ in a $3\mathscr{d}$ manifold $\mathcal{M}$. (If it helps we can then think of $\mathcal{B}$ as the magnetic field two-form and of $\star_h\mathcal{B}$ as the magnetic field co-vector.)

The one-form $\star_h\mathcal{B}$ then points in the direction of the positive normals of the surfaces with the largest $\mathcal{B}$-flux through them, and the modulus of $\star_h\mathcal{B}$ is given by the flux density of $\mathcal{B}$.

Let us make this a bit more precise: In $\mathscr{d} = 3$, a surface S has a normal direction which can be written in terms of a one-form field n_S, which has the property that

$$ \int_{\Gamma_\parallel} n_S = 0 \tag{B.11} $$

for any curve $\Gamma_\parallel$ in the surface S. This only specifies n_S up to multiplication of a scalar field, but we also define n_S such that it points in the positive direction of S. Then, n_S is defined up to a multiplication of a positive scalar field. (Since S and $\mathcal{M}$ have an orientation, we have a prescription of when a curve $\Gamma_\perp$, which intersects S at one point, passes through S in a positive direction (see Definition 9). We define n_S such that the integral

$$ \int_{\Gamma_\perp} n_S \tag{B.12} $$

is positive for short enough curves $\Gamma_\perp$, which pass through S in a positive direction.)

Since we have a metric, all surfaces have a well defined area and boundary length. If we now consider surfaces $\{s\}$ with sufficiently small area and boundary length, the integral $\int_s \mathcal{B}$ will only depend on the normal $n_s(X)$ of s at X and area(s), *i.e.,* the area of s. The direction of $\star_h\mathcal{B}$ is the positive normal to the surface with the largest value of $\int_s \mathcal{B}/\text{area}(s)$, and the modulus of $\star_h\mathcal{B}$ is given by $\int_s \mathcal{B}/\text{area}(s)$ for the same surface.

In Definition 10, we make it apply to forms of all orders. But, to do so, we first need to introduce the notion of a normal form of a submanifold.

Definition 9 *The normal form of a submanifold*
In a manifold $\mathcal{M}$ with dimension $\mathscr{d}$, a k-dimensional submanifold m has a normal form n_m of order $\mathscr{d} - k$. The normal form is defined such that

$$\int_{m_{\parallel}} n_m = 0 \tag{B.13}$$

for all $\mathscr{d} - k$ submanifolds $m_{\parallel}$, where either $m_{\parallel} \subset m$ or there exists an open subset $\mathcal{N}_m \subset m$ where $\mathcal{N}_m \subset m_{\parallel}$. The direction n_m is such that $\int_{m_{\perp}} n_m > 0$ for a $\mathscr{d} - k$ submanifold $m_{\perp}$ with an orientation such that the wedge product $V_m \wedge V_{m_{\perp}}$ between the volume form V_m of m and the volume form $V_{m_{\perp}}$ of $m_{\perp}$ is, up to a positive scalar field, the volume form of $\mathcal{M}$.

Definition 10 (*geometric definition*) *The Hodge isomorphism*
In a $\mathscr{d}$-dimensional oriented manifold equipped with a metric h, there is a natural map from k-forms to $\mathscr{d} - k$-forms, called the *Hodge isomorphism*. The image of a k-form a is called the *Hodge dual* and is denoted $\star_h a$.

Consider a set $S_{\epsilon,X}$ of k-dimensional, open balls parametrized by the ball radius ϵ and the center point X. By $B_{\epsilon,X}$ we denote a ball in $S_{\epsilon,X}$ with the property

$$\int_{B_{\epsilon,X}} a \geq \int_B a; \forall B \in S_{\epsilon,X}, \tag{B.14}$$

and by n_ϵ we denote the normal $\mathscr{d} - k$-form of $B_{\epsilon,X}$ that at X has the modulus $\int_{B_{\epsilon,X}} a/\text{area}(B_{\epsilon,X})$.

The Hodge dual $\star_h a$ is

$$\star_h a = \lim_{\epsilon \to 0} n_\epsilon. \tag{B.15}$$

Fact *Hodge dual and wedge products*

$$\int a \wedge \star_h b = \int \star_h a \wedge b \tag{B.16}$$

Fact *Inverse Hodge isomorphism*
If a is a k-form on a $\mathscr{d}$-dimensional manifold we have

$$\star_h \star_h a = (-1)^{k(\mathscr{d}-k)} a. \tag{B.17}$$

Fact *Hodge isomorphism and the Laplace operator*
The Laplace operator in a $\mathscr{d}$-dimensional manifold with metric h takes the form

$$(-1)^{\mathscr{d}+1} \star_h d \star_h d. \tag{B.18}$$

(To be clear, with Laplace operator we mean the operator which in Euclidian coordinates, i.e., where the metric has components $h_{\mu\nu} = \delta_{\mu\nu}$, takes the form $-\sum_\mu \partial_\mu \partial_\mu$.)

Definition 11 *Spatial slice*
A spatial slice $m(t)$ of space-time $\mathcal{M}$ is all points that correspond to a given time t.

Definition 12 *Pullback of a form*
Remember that we defined differential forms as maps from submanifolds to real numbers (e.g., a one-form is a map from curves to reals). Take two manifolds $\mathcal{M}$ and $\mathcal{N}$ and a smooth injective function ϕ from $\mathcal{M}$ to $\mathcal{N}$. Then, there is a natural mapping from the differential forms on $\mathcal{N}$ to the ones on $\mathcal{M}$, called the *pullback*. The pullback of a k-form a on $\mathcal{N}$ is given by the map

$$\mathcal{M}_k \to \int_{\phi(\mathcal{M}_k)} a \tag{B.19}$$

from k-dimensional submanifolds $\mathcal{M}_k$ in $\mathcal{M}$ to real numbers.

Definition 13 *Spatial isomorphism*
There is a natural isomorphism χ between all pairs of spatial slices. The isomorphism is defined in the following way: if a point X in $m(t)$ and Y in $m(t')$ is connected by an observer which stands still, then—when viewed as a map from $m(t)$ to $m(t')$—$\chi(X) = Y$.

This also provides an isomorphism between space-time differential forms and differential forms on spatial slices. The spatial slices are submanifolds of space-time $\mathcal{M}$, so there is the embedding map $m(t) \to \mathcal{M}$, which defines a pullback (see Definition 12) of forms in $\mathcal{M}$ to forms in $m(t)$.

All points in $\mathcal{M}$ are given by a doublet $(t, X(t))$. We can thus form a map from $\mathcal{M}$ to $m(t')$ which takes $(t, X(t))$ to $\chi(X(t))$, where χ now is interpreted as a map from $m(t)$ to $m(t')$. The pullback of this map provides a map from differential forms on the spatial slices to differential forms in space-time.

Definition 14 *Spatial form*
A spatial form is a form which maps back to itself when the spatial isomorphism is applied twice. With this isomorphism, these forms can both be viewed as forms on a spatial slices and on space-time.

When dealing with spatial forms, the spatial isomorphism will be kept implicit, but it should be clear when it is invoked.

Definition 15 *The spatial and temporal parts of a differential form*
With a notion of absolute rest we have a decomposition of any space-time k-form a into a two spatial forms: the spatial, $a^{sp.}$, and a temporal, a^t, part;

$$a = dt \wedge a^t + a^{sp.}. \tag{B.20}$$

Example 16 The electromagnetic field tensor
The spatial and temporal parts of the electromagnetic field tensor is the magnetic field two-form, $\mathcal{B}$, and electric field one-form $\mathcal{E}$, respectivly: $F = \mathcal{B} + dt \wedge \mathcal{E}$. The electric vector is defined by raising the index of $\mathcal{E}$ by the spatial metric h. The spatial Hodge dual of $\mathcal{B}$ is the magnetic field co-vector in $3\mathscr{d}$ and the magnetic field scalar in $2\mathscr{d}$.

Appendix C
Vector Bundles and Chern Numbers in Quantum Mechanics

Consider a subspace $\hbar(X)$ (of a Hilbert space $\mathcal{H}$) that varies continuously with the parameter X in some manifold $X \in \mathcal{M}$. The space

$$E = \bigcup_{X \in B} \{X\} \times h(X), \tag{C.1}$$

which is a subspace of $\mathcal{M} \times \mathcal{H}$, is called a *fiber bundle* and $\hbar(X)$ is called the *fiber* at X. When the fibers are complex vector spaces, as they are here, we use the more precise term: *complex vector bundle.* The mathematical structure of the fibre bundle can be described without the notion of the embedding space, $\mathcal{M} \times \mathcal{H}$, but for our purposes it is more practical to also use some of the extra structure that the embedding provides.

We now define the *Berry connection*, which we use to construct invariants that characterize complex vector bundles.

C.1 The Berry Connection

We start by picking a coordinate system $X \doteq \{x_\mu\}_{\mu=1,\dots,d}$ of $\mathcal{M}$, and a basis $\{|X; i\rangle\}_{i=1,\dots,N}$ that varies smoothly in some region of $\mathcal{M}$. We then consider a curve in E:

$$s \to (X(s), |s\rangle)\,; \qquad\qquad |s\rangle \in h(X(s)), \tag{C.2}$$

parametrized by s. We assume that the curve fits in the region of the bundle where the basis is defined, so we can write

$$|s\rangle = \sum_i \alpha_i(s)|X(s); i\rangle \tag{C.3}$$

T. Klein Kvorning, *Topological Quantum Matter*, Springer Theses,
https://doi.org/10.1007/978-3-319-96764-6

for some coefficients $\{\alpha_i\}_{i=1,\ldots,N}$. If $\hbar(X)$ is independent of X, there is no obstruction to having

$$0 = \frac{d}{ds}|s\rangle. \tag{C.4}$$

On the other hand, if $\hbar(X(s))$ varies with X, $|s\rangle$ has to change simply because $\hbar(X(s))$ does. The total Hilbert space has the usual notion of distance and provides a natural prescription for which vector in $\hbar(X(s+ds))$ is closest to $|s\rangle$, namely the linear projection of $|s\rangle$ onto $\hbar(X(s+ds))$. A curve that changes as little as possible is thus

$$|s+ds\rangle = P\left(X(s+ds)\right)|s\rangle. \tag{C.5}$$

Taking the inner product with $\langle X(s+ds;\alpha)|$ on both sides of the above equation gives

$$\begin{aligned}\alpha_i(s+ds) &= \sum_j \alpha_j(s)\langle X(s+ds); i|X(s); j\rangle \\ &= \sum_j \alpha_j(s)\left(\delta^{ij} - ds\left\langle X(s); i\left|\frac{d}{ds}\right|X(s); j\right\rangle\right) \\ &\equiv \sum_j \left(\delta^{ij} - i\left[\mathbb{A}_\mu\right]^{ij}\frac{\partial x^\mu}{\partial s}ds\right)\alpha_j(s)\ ,\end{aligned} \tag{C.6}$$

where we used $(\frac{d}{ds}\langle X(s); i|)|X(s); j\rangle = -\langle X(s); i|\frac{d}{ds}|X(s); j\rangle$. The last step defines the Berry connection,

$$[\mathbb{A}]^{ij} = -i\langle X; i|d|X; j\rangle\ , \tag{C.7}$$

where d denotes the exterior derivative in $\mathcal{M}$. (The symbol $\mathbb{A}$ (and $\mathbb{F}$ below) are matrices, and matrix multiplication is assumed. We use $[\mathbb{A}]^{ij}$ and $[\mathbb{F}]^{ij}$ to denote their components.)

C.2 The Berry Field Strength

We now want to consider basis independent properties of the Berry connection. The Berry connection relates coefficients of vectors in two different Hilbert spaces: one in $\hbar(X(s))$ and the other in $\hbar(X(s+ds))$. Thus, the matrix $\mathbb{A}$ can be taken arbitrary since we can change the basis of $\hbar(X(s))$ independently from $\hbar(X(s+ds))$.

By instead considering the smallest change along a closed curve, one gets an operator acting within a Hilbert space $\hbar(X)$. The explicit form of the matrix obtain in this way is basis dependent, but the trace of the matrix is not. We will now consider a curve that changes as little as possible around an infinitesimal closed path.

Subtracting $\alpha_j(s)$ and dividing by ds in (C.6) we can conclude that for a curve that changes as little as possible, the coefficients change according to

$$0 = \left(\frac{d}{ds} + i\mathbb{A}_\mu \frac{\partial x^\mu}{\partial s} \right) \boldsymbol{\alpha}(s), \tag{C.8}$$

where $\boldsymbol{\alpha} = (\alpha_1, \alpha_2, \dots)^T$. We now consider an infinitesimal loop gotten by moving the length dx^μ in the μ-direction, and the length dx^ν in the ν-direction, then backward in the μ-direction and finally back to where we started. For coefficients $\boldsymbol{\alpha}$ (that correspond to the smallest change along this curve) we have the equation

$$\begin{aligned} 0 &= \frac{1}{i}\Big(\left(\partial_\mu + i\mathbb{A}_\mu\right)\left(\partial_\nu + i\mathbb{A}_\nu\right) - \left(\partial_\nu + i\mathbb{A}_\nu\right)\left(\partial_\mu + i\mathbb{A}_\mu\right) \Big)\boldsymbol{\alpha}(X) \\ &= \left(\partial_\mu \mathbb{A}_\nu - \partial_\mu \mathbb{A}_\nu + i\mathbb{A}_\mu \mathbb{A}_\nu - i\mathbb{A}_\nu \mathbb{A}_\mu\right)\boldsymbol{\alpha}(X) \equiv \mathbb{F}_{\mu\nu}\boldsymbol{\alpha}(X)\,, \end{aligned} \tag{C.9}$$

where the last step defines the *Berry field strength*

$$\mathbb{F} = d\mathbb{A} + i\mathbb{A} \wedge \mathbb{A}. \tag{C.10}$$

We defined $\mathbb{F}$ specifically such that it has a basis independent trace, but it is instructive to see how this works by an explicit calculation. From the Definition (C.7) it follows, that under a coordinate transformation

$$|X; i\rangle \to \sum_j [\mathbb{U}]^{ij}\, |X; j\rangle, \tag{C.11}$$

the Berry connection transforms as

$$\mathbb{A} \to \mathbb{U}\mathbb{A}\mathbb{U}^\dagger - i\mathbb{U}^\dagger d\mathbb{U}\,, \tag{C.12}$$

while the field strength just rotates;

$$\mathbb{F} \to \mathbb{U}\mathbb{F}\mathbb{U}^\dagger. \tag{C.13}$$

Thus, because of the cyclic property of the trace, any product of the type

$$Tr\,(\mathbb{F} \wedge \mathbb{F} \cdots) \tag{C.14}$$

is basis independent.

Note that when taking the trace of the field strength, the second term,

$$\mathrm{Tr}\,(\mathbb{A} \wedge \mathbb{A})\,, \tag{C.15}$$

vanishes because of the cyclic property of the trace. We can therefore write

$$\mathrm{Tr}\left(\mathbb{F}_{\mu\nu}\right) = -i\sum_i \left(\frac{\partial}{\partial x^\mu}\left\langle X; i \left| \frac{\partial}{\partial x^\nu} \right| X; i\right\rangle - \frac{\partial}{\partial x^\nu}\left\langle X; i \left| \frac{\partial}{\partial x^\mu} \right| X; i\right\rangle\right). \quad (C.16)$$

C.3 Chern Numbers and Chern-Simons Invariants

We now have to consider the fact that there in general is no basis that varies smoothly over the entire $\mathcal{M}$. If $\mathcal{M}$ has a trivial topology (*i.e.*, that of an open ball) there is no obstruction to having a continuous basis that covers the entire $\mathcal{M}$. This is , however, not the case in general.

Let us start by studying some two-dimensional submanifold S of $\mathcal{M}$ and define

$$ch_1[\mathbb{F}](S) = \frac{1}{2\pi}\int_S \mathrm{Tr}\left(\mathbb{F}\right) . \quad (C.17)$$

This expression defines the first *Chern number*, ch_1, and the integrand (including the prefactor) is called the first *Chern character*. What we want to show in this section is that the first Chern number of a closed surface is an integer.

If S have a trivial topology, by using $\mathrm{Tr}(\mathbb{A} \wedge \mathbb{A}) = 0$ (which is true because of the cyclic property of the trace) we can write $\mathrm{Tr}\left(\mathbb{F}\right) = d\mathrm{Tr}\left(\mathbb{A}\right)$. This allows us to use Stokes theorem to rewrite the Chern number as

$$\frac{1}{2\pi}\int_{\partial S} \mathrm{Tr}\left(\mathbb{A}\right) . \quad (C.18)$$

Let us now calculate the first Chern number of a sphere[1] S. To evaluate the integral in (C.17), we partition the sphere into one region S_{big} that covers almost the full area, and a very small region S_{small} around the south pole. We assume that we can take the region S_{small} to be arbitrarily small, so that its contribution to the integral can be neglected. That is,

$$\frac{1}{2\pi}\int_S \mathrm{Tr}\left(\mathbb{F}\right) = \frac{1}{2\pi}\int_{S_{big}} \mathrm{Tr}\left(\mathbb{F}\right) . \quad (C.19)$$

Since the region S_{big} has a trivial topology, we can choose a basis that is smooth in the whole region and use Stokes theorem to get,

$$\frac{1}{2\pi}\int_S \mathrm{Tr}\left(\mathbb{F}\right) = \frac{1}{2\pi}\int_{S_{big}} \mathrm{Tr}\left(\mathbb{F}\right) = \frac{1}{2\pi}\int_{\partial S_{big}} \mathrm{Tr}\left(\mathbb{A}^{big}\right) , \quad (C.20)$$

[1]The statements we will prove hold for a general closed manifold. Considering e.g., the torus, one has to divide $\mathcal{M}$ into more pieces, making the proof a bit more involved. However, the arguments would be analogous.

where the superscript big means that $\mathbb{A}^{big}$ is defined with respect to a coordinate system that is continuous in the region S_{big}. We can, however, also write

$$\frac{1}{2\pi}\int_{\partial S_{small}} \mathrm{Tr}\left(\mathbb{A}^{small}\right) = -\frac{1}{2\pi}\int_{S_{small}} \mathrm{Tr}\left(\mathbb{F}\right) = 0\,. \tag{C.21}$$

In the overlap between S_{big} and S_{small}, both coordinate systems are continuous, and on ∂S_{big} they are related by some unitary transformation $\mathbb{U}$;

$$\mathrm{Tr}\left(\mathbb{A}^{big}\right) = \mathrm{Tr}\left(\mathbb{A}^{small}\right) - i\mathrm{Tr}\left(\mathbb{U}^{\dagger}d\mathbb{U}\right)\,. \tag{C.22}$$

Putting this together, we get

$$\frac{1}{2\pi}\int_{S} \mathrm{Tr}\left(\mathbb{F}\right) = \frac{1}{2\pi i}\int_{\partial S_{big}} \mathrm{Tr}\left(\mathbb{U}^{\dagger}d\mathbb{U}\right). \tag{C.23}$$

The matrix trace is the same in all bases, so we may consider it in the basis where $\mathbb{U}$ is diagonal,

$$\mathbb{U} = \mathrm{diag}\left(\xi_1, \xi_2, \ldots\right), \tag{C.24}$$

and we get

$$ch_1[\mathbb{F}](S) = \frac{1}{2\pi i}\int_{\partial S_{big}} \mathrm{Tr}\left(\mathbb{U}^{\dagger}d\mathbb{U}\right) = \frac{1}{2\pi i}\sum_i \int_{\partial S_{big}} \xi_i^* d\xi_i\,. \tag{C.25}$$

The ith term in this sum gives the change of the complex phase ξ^i accumulated when integrating over ∂S_{big}. Since ∂S_{big} is a closed curve, the phase change has to be a multiple of 2π, which proves that $ch_1(S)$ is an integer.

Reference

1. M. Nakahara, *Geometry, Topology and Physics*, 2nd edn. Graduate student series in physics (Taylor & Francis, 2003)

MIX
Papier aus verantwortungsvollen Quellen
Paper from responsible sources
FSC® C105338

If you have any concerns about our products,
you can contact us on
ProductSafety@springernature.com

In case Publisher is established outside the EU,
the EU authorized representative is:
Springer Nature Customer Service Center GmbH
Europaplatz 3, 69115 Heidelberg, Germany

Printed by Libri Plureos GmbH
in Hamburg, Germany